The Secret of Sherwood Forest

The Secret of

GUY H. WOODWARD
and
GRACE STEELE WOODWARD

Sherwood Forest

Oil Production in England
During World War II

UNIVERSITY OF OKLAHOMA PRESS : NORMAN

BY GUY H. WOODWARD

(with W. P. Z. German)
History of Conservation of Oil and Gas in Oklahoma and Kansas (Chicago, 1936)

BY GRACE STEELE WOODWARD

The Man Who Conquered Pain: A Biography of William Thomas Green Morton (Boston, 1962)
The Cherokees (Norman, 1963)
Pocahontas (Norman, 1969)

Library of Congress Cataloging in Publication Data
Woodward, Guy H.
    The Secret of Sherwood Forest.
    Bibliography: p.
    1.   Petroleum industry and trade—Great Britain.
I. Woodward, Grace Steele, joint author. II. Title.
HD9571.5.W66   338.2'7'2820942   72–12546
ISBN 0–8061–1094–5

Copyright 1973 by the University of Oklahoma Press, Publishing Division of the University. Composed and printed at Norman, Oklahoma, U.S.A., by the University of Oklahoma Press.

To the oil field "Roughnecks" who follow the drill bit around the world that mankind may have a better, longer, and more secure existence.

Preface

This amazing and hitherto untold story, born in secrecy, has remained buried in the private diaries, corporate files, and official records of government agencies for almost thirty years. The reason is not apparent. The purposes for searching out these facts and bringing them together in this book are two-fold. First, the events chronicled here constitute one of the most fascinating and important incidents of World War II; second, attention should be focused on the little-known, or at least seldom-realized, the all-important role that oil and oilmen played in the prosecution of the war. In the final analysis oil was indeed the key to victory of the Allies over the Axis powers.

War came upon Europe after eight months of diplomatic sparring that alternately raised the hopes of the people for peace or plunged them into despair and fear of war. Finally, the battle of words came to an end and World War II began. Giddy with his easy success in Poland, Hitler, on May 10, 1940, directed his well-trained and well-equipped panzer divisions, covered by the bomber and fighter planes of Göring's Luftwaffe across the borders of the peace-loving and defenseless nonbelligerent countries of Holland, Belgium, and Luxembourg. On the same day Winston Spencer Churchill became prime minister of Britain's government. Belgium surrendered unconditionally on May 25. The next day Holland and Luxembourg collapsed when their governments fled into exile. The seemingly invincible mechanized forces swiftly overran the Maginot line and swept into France.

The historic battle of Dunkirk reached its peak on May 30, and the retreat of 338,226 allied British and French troops appeared to be hopelessly cut off on the channel coast. But by a miracle of improvisation a hastily mustered fleet came to the rescue in the face of heavy and sustained air attacks. The fleet was made up of more than a thousand vessels consisting of everything from small fishing boats to ships of the Royal Navy. Despite this heroic effort, 68,111 men, killed, wounded, or taken prisoner were left on the beaches of Dunkirk.

On June 12 Paris was declared an open city and two days later, the German juggernaut, unopposed, rolled down empty avenues. Within a matter of hours the swastika was hoisted over public buildings. The French formally surrendered on June 22 at the same clearing in the forest north of Paris where Marshal Foch had dictated terms to the German Kaiser in November, 1918. To make the tragedy complete, the same railway car that had been used in 1918 was wheeled out of a museum to now serve as the funeral parlor for France's burial.

It was clear that England was next on the Führer's calling list. The British made ready for the ordeal they knew was upon them. A million and a half women and children were evacuated from the industrial cities to the countryside and to the villages on the west coast of Wales that offered more protection from the German bomber raids. Hospitals were made ready for the thousands they knew must be cared for.

On August 24, 1940, the first German bombs fell on London. By September 7 the Luftwaffe shifted to day-and-night bombing tactics and continued through fifty-seven days of around-the-clock bombing that reduced the city to rubble and will long be remembered in history as the London blitz.

Meanwhile, a far-seeing oil industry in the United States had perfected the miracle fuel known as 100-octane gasoline. During the first seven months of 1941, 681 vessels, including a large proportion of British and American oil tankers with their precious cargoes of aviation fuel and oil, went down in flaming seas. Some tankers came through and the young RAF pilots of Britain's Spitfire and Hurricane fighter planes received their first 100-octane gasoline in the spring of 1940. Hundred-octane gas gave Britain's air defense a new dimension of superiority in speed and maneuverability over the German bombers and fighter planes powered with the much lower octane synthetic fuels. With 100-octane gasoline, Hitler's forces were driven back across the English Channel, and Operation Sea Lion, Germany's code name for the Nazi's original plans to invade the British Isles, first in September and later in August, was definitely postponed. The Axis powers and the Allies alike were now fully aware that oil was the essence of victory. Those nations with access to oil would survive and win. By the same token those forces that were denied it were doomed to inevitable defeat.

After the fall of France, there was a growing feeling in America that Great Britain now stood alone between the United States and the Nazi's obvious bid for world domination. In January, 1941, the President called upon Congress to enact the Lend-Lease program and for the abandonment of the policy of neutrality and cash-and-carry. The proposal was introduced in Congress under the patriotic designation of HR 1776 and became law in March, 1941.

The oil industry was already engaged in an expansion of refinery facilities that was producing 40,000 barrels of 100-octane fuel per day.

World War II truly became a global affair at Pearl Harbor on Sunday, December 7, 1941. By August of 1942 oil was the prime objective on every battlefront throughout the world. Oil sparked every invasion. Without it no plane could fly; no tank could move; no ship could sail; no gun could fire. Oil shaped the course of battle on every beachhead, and from the enemy's lack of it, stemmed his inevitable and final defeat.

Early in 1941, realizing that he was engaged in a mechanized oil-fuel war, Hitler had frantically screamed into a microphone, "This war will be a long one. To fight it, we must be sure of oil for our war machine." On June 22 the Nazi armies commenced the greatest military campaign in history when they threw 257 divisions into the Russian front extending 2,000 miles from Stalingrad to the oil fields of the Black Sea basin. With the same determination to take the oil of the Middle East, General Erwin Rommel had in March, 1941, opened the first offensive in Africa and by 1941 had driven deep into Egypt to a position that caused *Time* magazine in August of 1942 to observe that Rommel now could see the Nile and had in his gun-sights the rich oil fields of the Middle East that lay beyond.

At the moment Prime Minister Churchill was in Moscow to explain why the long-hoped-for channel invasion of German-occupied Europe must be delayed in favor of an Allied landing in North Africa to save the Middle East oil from the rampaging armies of Rommel. Headquarters of the Supreme Command of the Allied forces had already been established in London. The whole of the British Isles had become an Allied arsenal of men, weapons, munitions, and supplies, first for the African invasion on November 11, 1942, and later for the invasion of the continent itself on D-Day, June 5, 1943.

The Japanese, following Pearl Harbor and their great need for oil if they were to become the masters of the eastern hemisphere, had caused the rapid penetration of the South Pacific by Japanese forces which overran and occupied the Dutch East Indies and Burma, two of the greatest oil-producing areas of the world. Meanwhile, the oil industry of America was meeting with incredible speed the growing military demands of the

United States and Great Britain, demands which by the end of 1942 had grown from 40,000 barrels of 100-octane fuel per day to over 150,000 barrels per day. Although the military and civilian demand of the United States and our Allies was a staggering challenge, the oil industry met the demands and piled up excess storage against the possibility of emergencies.

The tremendous quantities of oil needed for final victory was dramatized by the fact that it took 60,000 gallons of gasoline a day to keep a single armored division fighting. To keep the Air Forces operating a single day required fourteen times as much gasoline as was shipped to Europe for all purposes in the First World War. The bombing raids over Germany used up 10,000 gallons of toluene per minute. Toluene, a basic ingredient of TNT, gave the big blockbuster bombs battering Berlin their explosive power. To fill the fuel tanks of one battleship took enough oil to heat the average home for more than 500 years. One hour's flight of a Navy Hellcat fighter consumed enough gasoline to take the average automobile from Chicago to Los Angeles.

Oil is where you find it. Oilmen fought the war by following the drill bit, pipe line, test tube, and refinery plants, wherever they led throughout the world. Some fought the war in the frozen North at the Arctic Circle, others in the steaming, insect-infested jungles of the tropics. Others flew the hump over the Himalayas with fuel for our Chinese allies. Some spent their days and nights in the laboratories with their crucibles and test tubes. Many served their seven-day weeks as governmental aids in performing the many administrative duties demanded by total war. Thanks to a farseeing and wise nation that had refused to burden its oil industry with crushing controls by rigid laws, rules, and regulations, but had on the contrary encouraged the men engaged in the hazardous business of finding and producing oil.

Because of this policy, followed for more than twenty-five

years, the United States reserves were sufficient to sustain a growing, healthy economy in time of peace and now furnished security and the ingredients of victory in time of war.

It is against this backdrop of history that developed this important incident of World War II, perhaps the best kept secret of the war. The authors have enjoyed the research and the writing of this story. We sincerely hope the reader will find entertainment in reading it.

GUY H. WOODWARD
and
GRACE S. WOODWARD

Tulsa, Oklahoma
March 17, 1973

Acknowledgments

The authors of this book wish to thank the many thoughtful people who assisted us in gathering information for this book. During our research, both in the United States and in England, we were given great help and guidance by those who had either seen or had known the people we were writing about. And, although we are indebted to so large a number of individuals that we cannot name them all, we would like to acknowledge our special gratitude to the following: The officials of the Noble Drilling Company, Tulsa, Oklahoma, who released to us their hitherto secret *File E*; Mr. George Otey, Sr., of Ardmore, Oklahoma, attorney in 1942–43 for both the Noble and Fain-Porter Drilling Companies; Mr. Eugene Preston Rosser, superintendent for the English drilling project; Mr. Donald E. Walker, Rosser's assistant; the late Mr. Lloyd Noble, president of the Noble Drilling Corporation; the late Mr. Frank Porter, Oklahoma City, president of the Fain-Porter Company; Sir Philip Southwell, in 1942 manager of the Anglo-Iranian Company's Field Department, London; David Shepard, Petroleum Attache at the American Embassy; and Mr. James T. Murray, Technical Librarian, University of Tulsa, Tulsa, Oklahoma.

To Savoie Lottinville, Regents Professor Emeritus of the University of Oklahoma, Norman, we owe a special debt of gratitude. For it was he who first visualized the potentials of the great wealth of material we had acquired.

We also extend our humble thanks to our friends and to the members of our family who gave us patient understanding and encouragement during the years devoted to the research and writing of this book.

Contents

  XIX.   Farewell   240

          Bibliography   248

          Index   257

Illustrations

The Secret of Sherwood Forest

I.

The Need for Oil

German ground and air forces ushered in World War II by invading Poland on September 1, 1939. The principal nations opposing her by midsummer of 1942—from the beginning Great Britain, later Russia and the United States—could review a series of disastrous defeats.

The British recalled that long, hard time when they stood alone, were driven from the continent, their allies conquered. Eleven days before the formal capitulation of her ally, France, on June 21, 1940, Italy entered the war to join her Axis partner, Germany. These happenings presaged evil days for Britain: German U-boats from bases in France had easier access to the shipping lanes of the Atlantic, German airfields sprang up in the Low Countries and in France, and Italian submarines and ground forces posed a threat to the Egyptian oil fields as well as to oil coming from the Middle East.

The Axis powers took advantage of their strategic position. Because the Mediterranean was practically closed to British shipping, oil from the Persian Gulf had to be transported around Africa. German U-boats took a terrific toll along the sea-lanes of the Atlantic, as Adolph Hitler, the German Führer, launched his battle for Britain. One month after the fall of France, his forces began a campaign to sweep the English Channel clear of naval opposition, to destroy southern ports in England and all her airfields, petroleum storage facilities and installations, and finally, to reduce London to rubble by day and night bombing from the air. Foreseeing success in softening British resistance in this project, Hitler hoped to land sufficient forces in England in August or September for an easy conquest.

The first stages of his plan were put into effect in July. In late August air attacks were concentrated upon the London area and continued with lessening intensity into October, although by mid-September Hitler had abandoned plans for invasion. British Spitfire and Hurricane fighter planes could outmaneuver Luftwaffe bombers and fighter escorts, their course of flight picked up by English radar as they crossed the Channel. British planes were powered by an exotic fuel—100-octane gasoline—which enabled them to make fast take-offs, to accelerate more quickly, and to climb higher than the enemy. This fuel was a contributing factor to the victory won in the battle waged over three months' time and marked by Prime Minister Winston Churchill as Britain's finest hour.[1]

While this battle progressed, Japan on September 27 made a military alliance with Germany and Italy. This alliance recognized the leadership of Germany and Italy in the establishment of a new order in Europe as well as Japan's leadership for the same purpose in Greater East Asia. The Tripartite Pact was a warning to the United States that if she went to war with any of them, then the others would join against her. On March 11, 1941, Bulgaria became a party to the pact.

Hitler developed plans for a surprise attack upon Russia, although he had entered into a nonaggression pact with Russian dictator Joseph Stalin on August 23, 1939. Poland was to be partitioned with the proposed boundary between the two dictator states clearly delineated. Russia was to take over the Baltic states of Estonia, Latvia, and Lithuania and reclaim the

1 Sir Winston S. Churchill, *Their Finest Hour*, 301–98.

province of Bessarabia which had been lost to Romania in 1919. In return, Russia was to supply Germany with foodstuffs, oil, and other material when war was launched against Poland and her allies, France and Great Britain.

Suddenly and without warning on June 22, 1941, Hitler launched a massive attack upon Russia, a campaign which involved more men, material, and firepower than any in the history of wars. He expected to knock Russia out with one mighty blow. This he almost accomplished by capturing 3,800,000 soldiers and mountains of war supplies before wintry blasts slowed his offensive.

In his plans for action against Russia the following summer, Hitler advised his staff of the need for oil to win the war. Concentrated attacks were directed toward the source of 85 per cent of Russia's oil in the Black Sea and Caspian Sea areas, and Germany also sought to sever rail and water communications with industrial areas farther north. By August 8 Hitler had destroyed important pipe-line terminals, disrupted rail and barge traffic, and captured the Maikop-Kuban oil field which annually furnished Russia with 15,000,000 barrels of oil and which was only fifty miles from the main Soviet oil center near Grozny.[2]

Oil, or the lack of a source of supply, was a major factor in prompting Japan's attack upon the United States on December 7, 1941. Immediately after the Rome-Berlin-Tokyo Axis announcement in September, 1940, the United States denied Japan iron and steel scrap and aviation gasoline from this country, her principal source for these products. Japan continued to obtain crude oil until July, 1941, through an appeasement policy followed by our government in the hope this would divert the Japanese from overrunning the oil-rich Dutch East Indies.

[2] William L. Shirer, *The Rise and Fall of the Third Reich, a History of Nazi Germany*, 909, 914.

5

Japan called an additional million men to military service on July 2. She announced a protectorate over Indo-China on the 25th, and early in December had emissaries in Washington to negotiate on economic sanctions imposed by the United States.[3]

The Pearl Harbor attack seriously crippled the American battle fleet. Three days later, Japanese dive bombers sank two British battleships, the *Prince of Wales* and the *Repulse*, off the coast of Malaya. These actions gave Japan naval supremacy in the China Sea, the Pacific and Indian Oceans. By May, 1942, Japan controlled Northeast China, British Burma, and Hong Kong, all of Southeast Asia, the Dutch East Indies, Guam, and the Philippines with outposts on the islands of Attu and Kiska near the western extremity of the Aleutians.

The strategy of the United States for the war in the Pacific in the summer of 1942 was to try to keep the territory she held and to stockpile for major offensive actions in 1943–44. Anglo-American attention was directed to Fortress Europe and North Africa as well as toward the North Atlantic and the Mediterranean.

This did not mean there was to be no activity in the Pacific theater. The naval and air battles of the Coral Sea and Midway in May and June cost the Japanese four of their best carriers. These battles paved the way for the landing of the First Marine Division on August 7 on Guadalcanal, our first step on the long road back to the Philippines. American morale was considerably boosted by the April 18 raid on Tokyo by Colonel James H. Doolittle's squadron of B-25's from the carrier *Hornet*. His planes were fueled by 100-octane gasoline from the Anglo-Iranian Company's refinery at Abadan on the Persian Gulf.[4]

Adolph Hitler expressed a common belief when he said oil

[3] Dean Acheson, *Present at the Creation*, 19, 24–27.
[4] Henry Longhurst, *Adventure in Oil*, 106–108.

was necessary for victory. As stated in the official history of our petroleum administration during the war,

. . . oil did more than fuel and lubricate the ships and the airplanes and the motorized ground equipment. Oil was also heat and light and comfort and mercy. Out in the field, in the form of gasoline, it fueled the kitchens, it powered the radios and telephones, it warmed and illuminated the hospitals, it refrigerated the life-saving plasmas, it heated the instrument sterilizers, it ran signal devices, water purification systems, and repair machinery. From oil came the toluene for TNT that went into bombs, the asphalt for air fields, the jellied gasoline for flame throwers, the kerosene for smoke screens, the wax for packaging food and equipment, the petroleum coke for aluminum. More than 500 different petroleum products were regularly used by the armed services.[5]

Donald R. Knowlton, director of production in the Petroleum Administration for War, stated in 1943 that 65 per cent of the total tonnage of overseas shipping for the prosecution of the war consisted of petroleum products.[6]

Before the United States entered the war, a third of the world's oil production started to market, tanker-borne, over the waters of the Gulf of Mexico and the Caribbean Sea.[7] During the battle for Britain in 1940 Britain was sustained by oil from this area. The need increased for oil shipped from the Gulf Coast, as well as from Venezuela and Trinidad. In the summer of 1941 fifty American flag tankers carried petroleum and petroleum products from the area to Bayonne, New Jersey, and on to New York or Halifax for transfer to British tankers in an effort to reduce travel and turn-around time.[8] Tankers were bringing 1,400,000 barrels of crude oil or refined pro-

---

[5] John W. Frey and H. Chandler Ide (eds.), *A History of the Petroleum Administration for War*, 1.

[6] *World Petroleum* (January, 1944), 41.

[7] E. L. DeGolyer, "War Demands More Oil," *World Petroleum* (February, 1943), 26.

[8] W. Alton Jones, "Wartime Revolution in Petroleum Transportation," *World Petroleum* (April, 1943), 30–36.

ducts to the East Coast daily to meet domestic and British demands. By the end of May, 1942, delivery by tanker was down to a daily average of 121,000 barrels because of the success of German U-boats preying in wolf packs. From April to September, 1942, an average of one and one-half cargo ships a day were sunk in the Trinidad area.[9]

Plans had already been developed for a combined Anglo-American invasion of North Africa which was to occur November 8. American oil was desperately needed for the invasion and, as the chiefs of staffs worked out problems of logistics, the petroleum agencies of the two countries had the responsibility of meeting military needs for petroleum products. They maintained liaison representatives at meetings in London and in Washington.[10]

Hitler intensified his submarine war against Allied shipping. U-boats stationed in Norwegian fjords made the cold North Atlantic a graveyard for merchant vessels and seamen bound for Murmansk, and from the North Sea to the Bay of Biscay they came from their pens loaded with destruction to Britain-bound commerce. By the end of the summer the U-boats were sinking 700,000 tons of shipping a month—a tonnage in excess of what the shipyards of Britain and America were replacing.

With both German U-boats and the Italian ships operating there, the Mediterranean at the time was practically an Axis lake. The British-Iranian refinery at Haifa on the eastern

9 Philip Goodhart, *Fifty Ships That Saved the World*, 219. See also Frey and Ide (eds.), *A History of the Petroleum Administration for War*, 3, 87.

10 Government and industry worked closely together in Great Britain through the Oil Control Board, a subcommittee under the war cabinet, and the petroleum board, representing industry. An independent agency was created in the United States, the Petroleum Administration for War, hereafter referred to as PAW, which was manned in key positions by representatives of the oil industry. An advisory board composed of industry representatives, called the Petroleum Industry War Council, helped formulate national policies and programs and made recommendations to PAW.

shore of the sea was subject to bombing by the Italian air force, and Persian oil was diverted to Russia and the war in the Pacific.

The oil industry in the United States was operating under orders issued by the War Production Board, an emergency agency created in 1941. The agency was responsible for expediting government procurement of defense needs. It set priorities on scarce material in order that no segment of any industry necessary to the war effort should be discriminated against in the allocation of equipment or supplies.

The agency issued Conservation Order M-68 on December 23, 1941, to reduce the consumption of oil-field equipment and supplies during the next year to 60 per cent of that used in 1941.[11] In 1941, 29,574 wells were drilled; the following year, 17,872. Approximately 75 per cent of these numbers were development wells, drilled to lift oil from known reserves. After state regulatory agencies increased allowable production from proven fields, there was a minimal decline in daily production from 3,841,700 barrels to 3,795,800 barrels.

Some of this production, converted into petroleum products, was diverted necessarily to Pearl Harbor and the Southwest Pacific. If some of this supply, the policy makers decided, could be furnished from new sources, more oil would be available for the European theater. Drilling contracts were let to tap proven reserves in the Canadian province of Alberta. And, emphasizing the desperate plight to obtain oil for victory, General Brehon Somervell, chief of the United States Army Service Forces, on May 20, 1942, approved a contract for the Canol Project.[12] A small oil field had been found far to the frigid north near Norman Wells in the Northwest Territories of

11 DeGolyer, "War Demands More Oil," *World Petroleum* (February, 1943), 28.
12 *World Petroleum* (November, 1943), 29.

Canada. The contract provided for the drilling of additional wells and the building of a 500-mile long pipe line to Whitehorse near the Alaskan border. Any source of oil to aid the Allied cause was most welcome.

II.

Great Britain's Own Oil Fields

Britain's secretary of petroleum, Geoffrey Lloyd, called an emergency meeting in London of the Oil Control Board with members of the oil industry's advisory committee in mid-August of 1942. The purpose was to consider the impending crisis in oil. The Admiralty had reported fuel stocks were two million barrels below normal safety reserves and were sufficient to meet only two months' requirements. Reserves of approximately five million barrels were normally held in some forty widely scattered storage facilities. Bombing raids by the Luftwaffe had destroyed almost a million barrels in dock areas. At the same time increased military demands by the armed services further undercut reserves.

While he waited for the group to assemble, the secretary drew back the heavy blackout curtains to survey pensively the rubble partially blocking the nearby streets. The government buildings along Whitehall had not escaped. The gaping holes in their walls and rooftops seemed to hurl defiance at the madmen in Berlin. Some of the board members coming to the meeting, he felt sure, had joined thousands of other Londoners in spending the previous night in underground shelters. After the sirens sounded the air-raid alert, the underground railway tube always filled quickly.

Board members answering the emergency call heard and

discussed the secretary's reports of military and civilian needs including the very large petroleum requirements for the coming African campaign. The Supreme Allied Headquarters now established in London, the secretary said, would soon announce the date and plans for landing troops on the North African coast. At the moment, the secretary said he could reveal the fact that the prime minister was in Moscow to explain to the Russians that the across-channel invasion of France must be further delayed in favor of the African landings in order to save the Middle East oil from Rommel's rampaging forces. Following the discussion of this exciting news, the secretary called upon C. A. P. Southwell, an industry representative from the D'Arcy Exploration Company, to address the meeting on a matter he felt demanded immediate action.[1]

Southwell, a muscular, middle-aged man of medium height, had the clear blue eyes and ruddy complexion of a Britisher who spent much time out-of-doors. A good speaker, he quickly had the attention of his listeners. The development of Great Britain's own oil fields to their full potential, he told them, was the most pressing matter that should be immediately considered by those charged with the responsibility of furnishing England's petroleum requirements. With his statement concerning Great Britain's oil fields, he was at once loudly interrupted by his astounded colleagues: What oil fields? Where are the fields located? What is their production? Where is it being refined?

Southwell of course realized that only a few men in the meeting, except those associated with the Anglo-Iranian Oil Company, and its subsidiary the D'Arcy Exploration Company, knew of the discoveries of oil that had been made in the

[1] Southwell was a petroleum engineer of twenty years' experience with the D'Arcy Company, a subsidiary of the Anglo-Iranian Oil Company, Ltd., the world's largest oil company. He had served with the Royal Artillery in World War I and was awarded the military cross in 1918. See "Southwell, Sir (Charles Archibald) Philip," in *Who's Who 1971–72*, 2955.

Eakring area during 1939 and 1940. Calmly, he answered all questions directed at him, except one. He felt it unnecessary and even dangerous, he said, to pinpoint the location of the fields lest enemy intelligence pick up the information and direct an air attack upon the producing area. He explained that development had been going forward since early 1939. Presently, fifty producing wells had been completed in three areas at depths of 2,380 to 2,500 feet, yielding about seven hundred barrels of very high grade crude oil per day, an amount very small in terms of national war needs but still, a contribution. The producing area held, he felt, real possibilities.

The development had been slow, he said, for several reasons. The drilling equipment consisted of thirteen large rigs originally designed for deep-drilling operations at Anglo-Iranian locations in Persia. These drilling outfits with heavy 136-foot derricks, were not suitable for the rapid drilling in relatively shallow production. Too much time was required for building the derrick and moving the drilling equipment onto location and, after completion of the well, dismantling, removal to a new site, and re-erection of the derrick. At the present time it was requiring about nine weeks' operations for drilling and completion of a well and installing pumping equipment. Replacement parts for the well-worn rig equipment shipped home from Persia were in very short supply.

Southwell presented the recommendation of D'Arcy officials that an additional hundred wells drilled in the producing area would perhaps quadruple production. He stated that they had high hopes, too, that deeper drilling in other areas where promising geological information had been developed might lead to discovery of much larger producing fields. Since the time element was so important, Southwell proposed that a representative of the D'Arcy Company be given official authority by the secretary to proceed at once to the United States, where the most modern equipment was available, to purchase drilling rigs and equipment appropriate to develop the fields to their maximum potential. The equipment needed, he believed, could be acquired from manufacturers and suppliers in the United States. He must, however, it was pointed out, secure the approval and cooperation in Washington, D.C., of the Petroleum Administration for War and the Petroleum Industry War Council, the oil agencies having responsibility for furnishing the United States with its war requirements of petroleum.

Southwell forcefully closed his proposal with the important clincher that England's oil fields had double value, first, because they lay inland in a forested area safely beyond the enemy's submarine attacks, and second, because the area was blessed by a great natural camouflage covering of large matured trees that reduced to a minimum the danger of destruction by the dreaded Luftwaffe. The secretary expressed the unanimous opinion of the meeting that Mr. Southwell be furnished immediately with the official authority of the British government that might be needed to proceed at once to the United States for the purpose of acquiring the required oil-field drilling equipment to carry out the development program he had outlined to the secretary and his industry advisers.

After the meeting adjourned, Southwell took the late afternoon train to Grantham where he was met by his wife Mary for the short automobile trip to their wartime home, "Seven-Mile Post." During the train ride his mind had raced toward consequences of his proposal to the oil board and he had mentally reconstructed the events which led to the discovery of oil in Britain.

Before oil products came into commercial use during the last half of the nineteenth century, Britishers had used for various purposes seepages of petroleum, such as those found in Shropshire and near Formby in Lancashire. In the coal mines of Nottinghamshire the waxy green oil seepages were used for greasing axles of colliery wagons. The use of the seepage oil gained considerable popularity when it was found that oil-soaked peat made good fire lighters and the heavy tars formed by the seepage were excellent for caulking boats and for medicinal purposes on livestock. The first commercial efforts were developed with the discovery in 1850 by the Scotch chemist, Dr. James Young, that paraffin could be obtained by a crude method of distillation involving heating cannel coals and shale. The wax, thus obtained, proved valuable for making candles and sealing bungs of wooden casks used in aging and storing of beer, ale, and wines. Young's experiments led to the founding of the Scottish shale oil industry.[2] Many years were to elapse, however, before serious thought was given to the possibility that Britain might be an oil-producing region.

A stimulus in this direction was afforded during World War I when German submarines brought a threat to overseas supplies of essential petroleum products. At that time the government sponsored a search for oil, but results were disappointing. Only one well had found oil and even that was in such small quantity that it was abandoned as a noncommercial venture. One well in Scotland had shown signs of oil and some gas. Drilling of exploratory wells continued, but with an obvious decline of enthusiasm that petroleum in commercial quantities would be found in Great Britain.[3]

The significance of petroleum products to the world economy became increasingly important after World War I. Scientists developed more methods to substantiate hypotheses on

---

[2] C. A. P. Southwell, "Petroleum in England," *Journal of the Institute of Petroleum*, XXI (February, 1945), 27.

[3] *Ibid.*, 27–30. Francis T. Bradley's *The British Commonwealth and Its Petroleum*, an unpublished thesis in fulfillment of requirements for a Master's Degree, University of Tulsa, Tulsa, Oklahoma, goes into detail on the beginning of the petroleum industry in the British Isles.

the earth's substructure, methods to locate geological formations which trapped oil and gas. Experiments where there were known reserves indicated oil and gas in underground basins had some effect upon the magnetic and gravitational pull of the earth. As a result of these studies, the magnetometer and the gravity meter were developed. Experiments conducted, too, indicated that explosions set off underground produced shock waves which varied in intensity and time lapse as they traveled through formations of varying density. Out of these studies developed the seismograph and torsion balance.

Much geological field work was done in Great Britain in the nineteenth century because of the importance of coal to her economy. This had been continued under updated methods. George C. McGehee, a son-in-law of the well-known geologist-petroleum engineer, E. DeGolyer, while a graduate student at Oxford, from 1933 to 1935, had made a comprehensive geophysical survey of the British Isles. Southwell carefully studied these reports. His company had employed a number of young American geologists and geophysicists who were experienced in the use of gravitational, magnetic, and seismic methods in gathering data for mapping substructures.

The search for oil in Great Britain gathered momentum with the passing of the Petroleum Production Act of 1934. Although this legislation nationalized oil rights, it provided for the necessary granting of exploration licenses over substantial areas. By March, 1939, the D'Arcy Company held a block of 7,163 square miles and was carrying forward a more vigorous program of exploratory drilling than any of the other producing companies holding rights in Great Britain.

An analysis of data derived from coal exploration and workings indicated the prospect of an anticlinal structure in Nottinghamshire and adjacent counties. The structure was also indicated by seismic methods employed early in 1939 and subsequently by drilling later in the year when oil-producing sands were found in the Millstone Grit formation.[4] A combination of perseverance, confidence, and a willingness to take financial risks had finally brought about the important discovery by the D'Arcy Company of two producing fields. Then in 1940 geophysical survey teams and exploratory drilling in three nearby areas brought additional production.

Southwell, as he prepared to enplane for Washington, was confident. He felt sure that he could convince the Americans that they should make available the necessary drilling equipment to increase a safe and sure supply of oil for Britain in this moment of her great need.

[4] Alec H. Day, "Continuing Search for Oil in Britain," *World Petroleum,* Vol. XVII, No. 10 (September, 1946), 47–48. C. M. Adcock, "Three Decades of Drilling," *B P Shield,* London (March, 1968), 20–21. The *B P Shield* is a company publication of the Anglo-Iranian Oil Company.

III.

One-Man Mission to Washington

A night flight on Thursday, September 3, bound for New York by way of Montreal carried Southwell as one of its passengers. He had wanted the day flight out of London, but in these times of high pressure action and war restrictions one took what one could get, when one could get it, and was happy he had it. Ordinarily, considerably more time would have been required for obtaining a travel priority involving even vital war work. But in view of the growing government oil requirements for all phases of important military operations and in particular the impending Allied landing in North Africa, the petroleum secretary's office had seen to it that Southwell's request for air travel priority to the United States was promptly granted.

Settled for the flight, Southwell noticed for the first time that he was in seat thirteen.[1] "So what!" he thought; he had no time for silly superstitions. He was headed for a face-to-face showdown with the officials in Washington to obtain, by whatever proper method available, such oil-well drilling equipment that he and other production men of D'Arcy Exploration

1 Sir Philip Southwell recounted his experiences on this visit to America in interviews conducted by the authors in Oklahoma during 1968 and later at Great Yarmouth in England during 1969.

Company deemed essential to the speedy development of the known British oil reserves.

Southwell's brief case was carefully packed with maps, geological information, production data, engineering estimates, and other pertinent matter relevant to the British producing areas. The brief case was in handy reaching distance and never out of his sight. The entire world, it seemed, was crawling with enemy spies. Every passenger and even flight attendants were subjects of suspicion. Secrecy of those engaged in even the most remote as well as closely related war work had become, during the weary war years, an accepted practice. This was particularly true with Londoners. Southwell would soon find a similar prevailing attitude in New York and Washington.

The throbbing of the plane's engines seemed to beat out the warning, "You must not fail." Southwell reminded himself that now every drop of oil counted. He realized that at this juncture of the war when Allied forces were planning Operation Torch, the code name that had been assigned by the military to the approaching invasion of North Africa, that petroleum was the key to victory.

Sleep on the flight was fitful because his restless mind was mulling over the approach he should make to Washington officials. His presentation must be complete, effective, and compelling. Obtaining Washington's help in securing the drilling equipment that was required for rapid development of the British oil fields was not going to be an easy task since such equipment was already in short supply in the United States. At last Southwell was aroused from his half sleep by the announcement over the loud speaker that the plane would be landing in fifteen minutes.

Suddenly the Clipper broke out of the morning fog over Montreal's airport where European passengers to New York first landed in America. The big plane hit the lighted runway that loomed shadowy in the early morning dawn. He must hurry. Knowing that he had only a few minutes to board the connecting flight to New York made him unduly impatient at the time consumed by customs officials. Customs cleared, Southwell felt relieved. Entering the New York plane, he checked to see if the brief case was secure. The precious brief case clutched under his right arm he found was reassuringly locked. He would have preferred that the contents had not been revealed even to customs officials, but war was war, and he realized that the tight security was, of course, necessary.

B. R. Jackson, attorney and representative of Anglo-Iranian Oil Company, Ltd., in New York, had cabled him before departure from London that he and Cartwright Reid, a young Britisher in Jackson's office, would meet him at the New York airport and from there the three would proceed to Washington. A late afternoon appointment had already been arranged for Southwell and Reid with Mr. Don Knowlton in Washington. Knowlton was the deputy administrator in charge of PAW's production division.

Finally at 4:30 P.M., Southwell, having arrived in Washington, immediately became a part of what seemed to be the never-ending and confused groups moving in and out of the PAW offices now housed in the east wing of the Department of the Interior building.

The heat and humidity of Washington were oppressive. But after a few minutes, the trim, good-looking young lady at the information desk motioned to him and said that he might now see Mr. Knowlton in Room 2020. Southwell made a mental note of the wartime practice of the receptionist entering his name and the time of his visit on the large sheet of ruled paper that lay before her, as he and Reid were ushered into Knowlton's office.

Introductions were brief and small talk of Southwell's flight was concluded in short order. Instantly, Southwell and Knowlton were friends with common objectives of winning a war that seemed to become more menacing by the hour. They had also the very important common background of well-informed and experienced petroleum engineers. Southwell had previously decided and Reid had agreed that a brief, succinct statement of his business in Washington would be the best approach. The people in PAW seemed to be moving at an accelerated speed. Department heads and other personnel moving in and out of the office, it appeared, had little time for talk except concerning matters of essential importance.

So it was that Southwell, with the briefest of reference to the dire needs of his government for petroleum supplies, told Knowlton he was in Washington for the purpose of acquiring ten of the latest model rotary oil-well drilling rigs suitable for drilling to depths of twenty-five hundred feet. Two strings of three-and-one-half-inch and two strings of four-and-one-half-inch drill pipe and a supply of rotary rock bits would be required. The rigs, Southwell explained, were to supplement old, oversized and cumbersome equipment now being used in the development of Britain's oil fields. Knowlton was obviously surprised. He did not know that oil was being produced in Britain and said as much.

"Few people do," Southwell replied, "and such fact will not be revealed if we can prevent it to anyone other than those involved in the operations. The critical situation in Britain with respect to oil is such that if the Nazis knew the location of our production, they would not rest until every well had been bombed or sabotaged. Already, a few German paratroopers have been picked up in the area of the oil fields. We just hope that we have been lucky enough to get all of them." Then, with considerable emphasis and looking directly at the deputy administrator, he said, "Mr. Knowlton, I trust you understand that secrecy in this matter is most important. We should like very much for the talks with the people we see here, to be absolutely confidential." "I understand," Knowlton assured Southwell. At the end of the conference Knowlton agreed to meet Southwell again on the following Monday, a holiday in the United States. Knowlton explained that Monday was Labor Day but that he would be in his office. The presence of war seemed very close to Southwell.

Knowlton already knew that the British petroleum position was grave from reports just received by PAW officials from London. Since May, David Shepard, petroleum attaché for the Department of State with offices at the American Embassy, had been accorded membership on the Oil Control Board at the request of the secretary for petroleum.

Knowlton also had to weigh the effect that approval of Southwell's request might have on the home front, operating as it was under wartime restrictions. Only a month before, on August 3, work had begun on the 24-inch pipe line known as the "Big Inch," a 1,254-mile project from Longview, Texas, to Philadelphia and the East Coast. The system would hold at capacity 3,836,000 barrels of oil—a perpetual reservoir of oil moving constantly eastward. This had intensified the competition for steel for other emergency projects. The rapidly increasing demands for new petroleum synthetics—100-octane aviation gasoline, toluene for explosives, and raw materials for synthetic rubber—led to the use of critical materials in the modification and expansion of the refining industry.[2] His mind dwelt on the importance to the air force of 100-octane gasoline and recalled with satisfaction that industry had taken the lead in providing quantities needed, even in excess of government estimates.

The British, Knowlton recalled, had produced the superior aviation gasoline in small quantities through an alkalization

[2] William R. Boyd, Jr., "Oil Industry Moves Deeper Into War," *World Petroleum*, Vol. XIV (January, 1943), 22. Boyd was president of the American Petroleum Institute and during the war served as chairman of the Petroleum Industry War Council.

process at their Persian Gulf refinery. But needing equipment to increase their output, they had requisitioned and obtained from the United States a complete catalytic cracking unit and superfractionating towers for installation at Abadan in mid-1941. The material was dispatched in three British merchant ships traveling in separate convoys. All were sunk and the supplies lost. Now only recently a convoy had made delivery of other parts to erect units at the British Abadan refinery to produce greater quantities of aviation gasoline so essential for the war in the Pacific.[3]

Knowlton thought, too, of how rationing was affecting American families. First, the sale of automobile tires was frozen on January 5, 1942, and steps were established by the Office of Price Administration to ration them according to essentiality. Recapped tires and recapping services followed in February. Automobile rationing appeared in March, along with processed foods, meat, and sugar. He had heard a Washington rumor that gasoline rationing would be extended into the oil-producing southwestern states in November, and he smiled wryly at the thought of what the reaction to this would be by some of his Texas-Oklahoma friends.

Southwell was right. Yes, it was a good idea that no publicity should accompany his proposal to transfer American drilling equipment to Britain.

Southwell met with Knowlton on Labor Day and the following morning with PAW attorneys who promised an opinion before the end of the week. On Thursday Southwell learned that priorities to purchase the equipment, such as he felt he must have, could not be issued to foreign corporations, a foreign citizen, or to a foreign government. Some other method had to be developed. Southwell had in discussions with Basil Jackson and Cartwright Reid considered even the possibility of a lend-lease arrangement. But lend-lease would require a

3 Longhurst, *Adventure in Oil,* 107.

new start with new people and a completely different approach. The Lend-Lease Act had been passed by Congress in late March of 1941 as a means to aid Britain and later was extended to Russia and other Allies.[4] Lend-lease would require negotiations between the governments of the United States and Great Britain and would undoubtedly cause additional long and indefinite delays.

During his talks with Knowlton, Southwell had been developing a new idea. The more he thought of it, the more certain he was that it was feasible and, with the cooperation of PAW, would work. On September 11 at 10:00 A.M. by appointment the group assembled in Knowlton's office. Southwell at once presented his idea to the group. Since D'Arcy Exploration Company, he reasoned, could not be given priority for purchase of the required equipment, there was no legal reason why a drilling contractor, a citizen of the United States, or a United States corporation engaged in the business of oil-well drilling on contract, could not be employed by D'Arcy to conduct the contemplated operations in the United Kingdom and the contractor be given a priority to purchase the drilling rigs and other necessary equipment. Knowlton and J. Howard Marshall, chief counsel for PAW, considered this approach. To Southwell's great elation, Marshall said that he thought such an arrangement would be wholly within the law and regulations. In fact there were precedents for such procedure. United States drilling contractors operating in Canada were using United States government priorities for the purchase of needed material and equipment. Before adjournment, it was agreed that Knowlton would contact several contractors and arrange a meeting between them and Southwell. Knowlton was to advise Southwell as soon as he could make the appointments.

Much to the satisfaction and relief of Southwell, a telephone call from Knowlton the next day advised that two California

[4] Acheson, *Present at the Creation*, 27–33.

drilling contractors and two from Oklahoma had been contacted. They would meet with Southwell in Knowlton's office the following Monday. Accordingly, two contractors from California, Mr. Lloyd Noble of Ardmore, Oklahoma, president of the Noble Drilling Corporation, and his right-hand man Mr. C. C. Forbes, vice-president of the Noble Corporation, with headquarters in Tulsa, Oklahoma, and Mr. Frank Porter, president of the Fain-Porter Drilling Company with offices in Oklahoma City, arrived at Knowlton's office as scheduled. Soon after the arrival and a brief discussion of the matter, the gentlemen from California explained that it would be utterly impossible for them to consider additional drilling operations because of previous commitments, and immediately withdrew from the meeting.

Noble and Porter remained to discuss the matter at length with Southwell who explained his program. Both men listened attentively, asking many questions to develop a clear picture of the D'Arcy Company program. Southwell, however, was careful not to divulge the location of the oil fields involved. He merely said that they were within the United Kingdom. Although reluctant to turn Southwell down in view of the critical need of Great Britain for oil, the two Oklahomans finally said they did not feel they could take on this added responsibility because of previous commitments.

Mr. Noble pointed out that he had just concluded an agreement with the United States government to drill a number of wells in the Northwest Territories of Canada near the Arctic Circle, known as the "Canol Project." Here was being developed a crude oil supply for a refinery at White Horse, Canada, from which petroleum products would be shipped to the Pacific theater and possibly to Russia.

Mr. Porter felt that his personnel and operations were too small to take on the work such as that outlined by Southwell. Eventually, both Noble and Porter left. Noble said he was flying to Ardmore that night because of the many matters awaiting his attention. He boarded the company's plane and the pilot flew him directly to Ardmore, where they arrived in the early hours of the morning.

When Noble and Porter had gone, Southwell felt at sea again. The high hopes that he was on the way to action were dashed at the refusal of the contractors to consider his project. Yet Southwell, with the tenacity that so definitely characterized the British war leadership, was not to be put down with one refusal. He thought that he detected a soft spot in Noble's attitude; he must press the matter further. Maybe he had not made the urgency of the situation completely clear. Southwell, unable to sleep, paced the floor of his hotel room. He had been deeply impressed by Noble's obvious experience and his quiet assurance. Noble was the man he wanted. The Britisher decided that he would go at once to Ardmore while the conference was still fresh in both their minds.

He called Cartwright Reid to see what could be done about a seat on the next plane to Dallas, the nearest airline service to Ardmore, Oklahoma. In a few minutes Southwell's telephone rang, and Reid told him that with the help of the British Embassy, arrangements had been made for him to hitchhike a ride to New Orleans on a British naval plane that was leaving from Andrews Air Force Base, a few miles out of Washington, at 5:00 A.M. Hurriedly throwing a few things into his bag, Southwell took a taxi to Andrews field and arrived at the airport in time to identify himself and get aboard the waiting plane.

His arrival in New Orleans soon after daybreak enabled Southwell to apply for early morning travel by plane to Dallas. However, getting a seat proved difficult, since the plane was crowded and he had no priority for travel on commercial lines. Southwell was told to stand by. At the last minute he was called to the counter and ticketed. At Dallas a rental car was available.

Signing a card stating the rental car would be used exclusively for business, he received one tank of gasoline and was on his way. What he would do when that one tank of petrol was gone did not at the moment disturb Southwell. It was near 10:00 A.M. when Southwell drove into Ardmore, Oklahoma.

After a few directions from the people on the streets, he located Noble's home and pressed the button at the front door. Noble came to the door in pajamas, sleepy-eyed, and hair on end. His looks showed plainly how tired he was. Obviously surprised, he invited Southwell in and asked how it happened that he was in Ardmore at that time of the morning. Briefly, Southwell sketched his wild trip to Ardmore via New Orleans and Dallas.

Noble was undoubtedly impressed by Southwell's efforts. The guest was invited back to Noble's bedroom where conversation continued from bedroom to bathroom while Noble shaved and showered. Noble explained that his wife had passed away in 1936, and he and their children, Sam, Ed, and Ann with the aid of a housekeeper kept the home.

The Britisher and the native Oklahoman who was born in the Chickasaw Nation, were drawn together by common experiences. Both had served in World War I—Southwell with the Royal Artillery; Noble in the American Navy. Both were college men—Southwell with a degree in petroleum engineering from the University of Birmingham, while Noble left the University of Oklahoma after completing his freshman year in the School of Law. Both had been part of the tremendous advances made in the oil industry since the 'teens; Southwell in the British Empire; Noble from the time he was a youthful roustabout in the Healdton Field before war service was presently operating as a drilling contractor from southern Louisiana to Montana and from southwest Texas into Illinois.

Curious about the background of this modest man, whose name was known wherever there was drilling for oil, Southwell asked about the Indian governments before statehood. Noble explained the conditions under which his parents had been permitted to settle in Pickens County, Chickasaw Nation, now Carter County, Oklahoma, and how, before statehood and public schools, he attended Hargrove College, a Methodist elementary school where many of his classmates were members of the Chickasaw tribe. He elaborated on the long, thin line of Indian heritage which made Oklahoma unique. He pointed out that a former intermarried citizen of the Chickasaws, Governor William H. Murray, had appointed him to the Board of Regents of the University of Oklahoma, a position to which he had recently been reappointed and chosen as the presiding officer. Finally, Noble said casually but choosing his words carefully that if "Porter would join me, I would be willing to get together again and see if we can work out a contract to purchase the equipment for D'Arcy and see if we could recruit enough men to run the rigs." He added that he would not consider the job under any circumstances if it were not a matter so essential to the war effort. Noble then surprised Southwell by telling him that he would not expect to take any profit out of the work. After reimbursement of cost and expenses and overhead, the work, he said, would be a contribution to winning the war by Noble Drilling Corporation and the Fain-Porter Company. The work getting under way in Canada near the Arctic Circle was, he said, on the same basis. As Noble dressed, they talked. A number of telephone calls interrupted the conversation as Southwell attempted again to give Noble a picture of the desperate plight of the British for oil and the extreme importance of rapid development of Britain's maximum potential oil production.

Finally, Noble got around to calling Porter. After considerable talk, Noble obtained Porter's agreement to meet with Southwell and discuss further the possibilities of Noble and Porter taking on the job. It was agreed the group would meet later in the week with Mr. Knowlton in Washington. Noble pointed out that obtaining equipment and recruiting men would be wholly dependent on priorities and the approval of United States government authorities, and Knowlton would be the man who must clear the way for the contractors. Southwell, tired but encouraged, immediately returned to Dallas, after getting Noble's aid to purchase a tank of gasoline, and boarded the first plane possible for New York and a day and night of much needed rest.

A few days later, Mr. Ed Holt and P. M. Johns of the Noble Corporation, Frank Porter of the Fain-Porter Drilling Company, and Mr. George Otey of Ardmore, Oklahoma, an attorney who represented both Noble and Fain-Porter, met with Southwell, Jackson, and Cartwright Reid. Mr. Ralph Shulthius and Mr. George Walden of the PAW staff in Washington were present. Holt said that Noble and Forbes could not be present because of the press of other matters. Southwell was particularly impressed by the comment of George Walden that these oil wells located inland in Britain were really "unsinkable tankers," far out of the reach of the active German submarine packs.

Southwell outlined for the group his company's program for the completion of at least an additional one hundred wells during the next twelve months. As the discussion developed, Holt observed that ten rigs were more than would be needed. He said that four rigs, he felt, could accomplish the job. Holt suggested that since the amount and type of equipment must first be decided on, that Southwell go with him on a quick trip through the southern Illinois fields where considerable drilling was being conducted with equipment similar to what he believed would be appropriate for the British project.

The meeting adjourned to meet again at Jackson's office in New York the following week. In the meantime, Holt and Southwell toured the Illinois fields where Southwell was to see for the first time the new type of unitized rigs, equipped with eighty-seven-foot jackknife masts instead of the conventional wooden derricks. Southwell seemed convinced that the equipment he saw was what was needed. With more hope than belief, Southwell reluctantly agreed with Holt that probably four rigs would be sufficient to do the work.

Although Southwell was well pleased with Noble and Porter and their company men, he felt that a contract covering the details agreeable to the parties must be worked out. He had been attracted from the first to both men. Noble was a muscular man of medium height with unruly blond hair and a tanned face that gave every evidence of much time spent outdoors. His penetrating blue eyes looked one squarely in the face. The wrinkles at the temples somehow made you know that there was kindness and understanding behind the intelligent face. Noble's casual dress bespoke a man equally at home in the field and in the office, and the habit of nibbling an unlighted cigar telegraphed his concentration.

Porter was a somewhat larger man with the unmistakable face of a second generation Irish-American, in fact the son of a Brooklyn policeman. His pink complexion and quality attire reflected a successful businessman of ability and determination. Both men radiated a quiet assurance. The premeditated, almost hesitant, answers to Southwell's questions made him conscious of the integrity and forthrightness of the Noble and Porter men. Southwell had the good feeling that, at last, he had the right people. In fact, he told himself, he could actually leave the terms of the contract to them. He knew one thing for sure: the D'Arcy Company and the war-weary people at home were fortunate to have men like Noble and Porter on their side. He could say as much for the other Noble and Porter associates he had met as well as Knowlton and all the PAW personnel. He was now in a position to assure his company officials and the Secretary of Petroleum that development of the British oil fields would move forward as soon as contractual arrangements could be completed to have men and equipment on the site.

George Otey, the attorney representing both the Noble and Fain-Porter companies, suggested a few matters that must be covered by the contract.

Otey's southwestern twang, white Stetson, bright shirt, string tie, and embossed western boots attracted the attention of busy New York, but to Otey's delight, gave no indication that he was considered one of the nation's outstanding oil and gas lawyers. Otey's dress and speech indeed furnished a sharp contrast to the Britishers' muted business attire and their clipped Oxford accents. But as the days of negotiations stretched into weeks, Jackson and his assistants were to recognize Otey's great legal abilities and his sympathetic understanding of the many complex problems that must be covered by the contract. These included (1) deciding on the make-up and number of drilling crews needed; (2) the finding and employment of personnel; (3) transportation; (4) draft board deferments; (5) health certificates; (6) passports; (7) official priorities for the purchase and delivery of drilling equipment; (8) insurance; (9) reporting and accounting procedures; and (10) the many and various other items that necessarily would be involved. Compliance with the numerous governmental rules and regulations, Otey frequently pointed out, would require time and patience.

It was now near the end of October, 1942. Almost two months of continuous discussions had followed the group's first meeting, sometimes continuing far into the chill autumn evenings that had descended on New York City. It was clear that even now considerably more time would be needed. Southwell was growing even more impatient and anxious to return home as soon as possible. Many things in England were demanding his personal attention.

Finally, after many meetings through September and October, Southwell decided he could wait no longer and made application for air passage to London. B. R. Jackson and Cartwright Reid were left to come to terms on the many details of the contract. Major issues, he felt, had been settled. The agreement would constitute, in effect, a work contract between D'Arcy and the Noble and Porter companies. Noble and Porter would purchase the equipment for the account of the Anglo-Iranian Company pursuant to priorities which the PAW had agreed to issue; they would employ the drilling crews, truck drivers, welders, mud men, and other technical assistants including one over-all supervisor and one assistant. The personnel would be employees of the Noble company which would act in performance of the contract for both the Noble and Porter companies.[5]

Before leaving New York, however, Southwell got off a letter on October 30, 1942, to Lloyd Noble at Ardmore, Oklahoma, in which he advised:

I am waiting to get away and hope I shall not be delayed. I am sorry we could not meet again before I left. However, with Mr. Otey and Mr. Holt in New York it has been possible to get matters to a stage which to the best of my belief is satisfactory to all parties.[6]

[5] "George Otey to Lloyd Noble," October 21, 1942, *O* file; "C. A. P. Southwell to Lloyd Noble," October 30, 1942, *E* file. A separate confidential file was maintained on the "English Project" with access restricted to the few who had need to know.
[6] "Southwell to Noble," *E* file.

IV.

"Mr. Rosser, Meet Mr. Walker"

Monday, November 9, 1942, was an overcast day in Casper, Wyoming. Eugene Preston Rosser, the new assistant to Marion Tate, general superintendent of the Rocky Mountain division of Noble Drilling Corporation, arose early, as was his custom. He had a coffee date with some of the Stanolind boys (Stanolind Oil and Gas Company) at the coffee shop across the street from the Noble offices. He slipped out noiselessly because his wife Bernice and young daughter were not yet up. He had taken the drilling report from the men in the field, which had given him a run-down as to depth, condition of the hole, repair, maintenance, supplies needed at each rig, and other items of interest. The fishing job at Elk Basin had been cleaned up and the crew was drilling ahead. It was a good feeling to know that there was no well trouble at the moment.

As he backed his car out of the driveway, he heard the honking of Canadian geese traveling in a flying wedge southward to winter havens in Arkansas and Louisiana. "No hunting for me this year," the thirty-year-old Rosser thought, as he watched the geese overhead. Too much demand for drilling contractors. The big boys in Washington were certainly putting on the pressure for new drilling and exploration. This was particularly true with respect to the development of proven oil reserves.

Stanolind was now contemplating the need for two additional rigs within a few days. Just how Noble could spread its rigs to satisfy the growing customer demand was becoming a problem. In fact, the situation was already tight. Mr. Tate had warned Rosser to be careful about making commitments until they could determine exactly the number of drilling rigs they would be able to assemble with the equipment on hand and what new equipment they could expect to get delivered and when they might receive it. This damn war, Rosser thought, had thrown everything out of joint.

Drill pipe was becoming the big problem. Drilling mud, too, was rationed by the supply houses because tannic acid, an ingredient, was needed for the tanning and treatment of leather. The Tulsa office was doing a pretty good job in getting drill pipe, but even so, delivery was slow and delivery dates were rarely met. Rosser thought that equipment and supplies would be forthcoming if someone could convince those "birds" in Washington that the development of additional oil production was really important to the war effort. Where did these "silly" bureaucrats come from that had been sucked into government jobs since Pearl Harbor? Where did they think that additional oil was coming from unless more wells were drilled?

At the coffee shop, the Stanolind boys were already gathered in a booth. With morning greetings, Rosser joined them for coffee and a sweet roll. Conversation concerning the two new wells Stanolind wanted to get and the type of equipment Noble could furnish had only begun, when the telephone at the cashier's desk rang. Rosser was summoned to the telephone. Mr. Tate's secretary informed him Tulsa was calling, and Mr. Tate thought it best for Rosser to take the call in his office. "What's up now?" thought Rosser as he excused himself from the group and hurried across the street and upstairs to take the call. Impatiently he waited for the operator to get the calling party on the line. In a few moments came the clipped, well-known voice of Ed Holt, vice-president and general superintendent of the Noble Drilling Corporation.

"Gene," he said, "Lloyd wants you to come to Tulsa right away. He says pack up everything, bring Bernice and the baby, and be in Tulsa Friday morning." That meant he had only four days to help with the inventory of equipment he and Mr. Tate were working on and make arrangements for leaving, do the packing, and see the rationing board to get gas stamps. If they were going to take wearing apparel and the baby's things, they must have the car. Train travel was difficult since every Tom, Dick, and Harry seemed to be engaged in some kind of war work somewhere.

But, if that is what the boss wanted, he would make it. Rosser's career had been one of doing what had to be done, when and where it was necessary. If the boss says pack a bag and live out of it, he grumbled to himself, why you pack a bag and make things do if it takes a month's time. In the rough, tough, competitive times of the depression thirties he had survived, and that was something—survived the hard physical labor demanded of roustabouts and roughnecks, the unpalatable food served in shantytowns, the unrestrained rowdyism of oil field workers at play—not only had he survived but he had mastered all jobs through holding and doing every job connected with the drilling of oil wells.

Rosser himself was a product of this part of the industry. He had hired out to the Noble Company at an early age under the pretense of being a boilerman, after being told by a friend who was eavesdropping on a telephone line that Red Smith, the driller at a certain location, was in need of a potboiler man. With this meager information, Rosser had hurried to the location in the middle of the night. When Red Smith, the driller, showed up for the morning tour, Rosser, "by working like hell," as he recalled, had up a good head of steam in the

35

potboiler that sat near the doghouse at the end of the runway of the old cable tool rig. The regular boilerman, who had gotten "bagged" at Mabel's place, never showed up again, and Rosser, by fudging a few years on his age, went on the Noble payroll. He had never since that time nor would he ever in the future, until retirement, know any other economic affiliation.

Rosser remembers he almost lost his big opportunity to become a part of the oil industry when Red found that Rosser had dumped into the slush pit a five-gallon can of dirty water he had found on the boiler, in which a pair of Red's greasy overalls were being soaked with the help of half a can of lye. Red was not fully appeased until Rosser had fished out the overalls, rinsed them with hot water from the boiler, and spread them on the roof of the doghouse to be dried by the sun.

On that November day in 1942, Gene immediately called his wife to tell of his summons to report to headquarters. Three days later they left Casper. Rosser drove to Denver with Bernice and the baby, and then his wife drove on to Alvin, Texas, to visit her mother while he rode the train to Tulsa. He arrived Friday morning, November 13, to find Holt waiting for him with Whitey Stark, one of Noble's top truck drivers and an expert on trucks and heavy moving equipment.

When he asked, "What's the deal?" Holt told him Lloyd could not be in Tulsa but wanted to meet with him in Ardmore the following day.[1]

"What the hell does he want to see me about?" Rosser wanted to know.

"That's something Lloyd will tell you about tomorrow. He has the details. If you'll wait here a little while, we have a job to attend to today."

[1] Interoffice memoranda, records of telephone calls and correspondence filed in the *E* file, Noble Drilling Corporation, are the basis of dates cited and the substance of transactions herein recorded. These official records were supplemented through taped interviews with the authors recorded by Eugene P. Rosser and Donald E. Walker at various times, 1968–70.

For the next hour Rosser moved around the office, visiting with friends, reading the morning paper, but no one gave any indication of why he had been called to Tulsa. He found out later that except for Holt and two or three others in the office no one else knew the reason.

Holt, moving at his usual rapid speed, called to Gene down the hall, "Come on. Let's go."

"Where the hell are we going?" Gene asked.

"You and I are going down to the International Harvester place with Whitey and select some trucks and winch equipment to go with four rigs we are sending to a foreign country." For the first time since leaving Casper, Rosser was given a hint of the reason for his trip to Tulsa and Ardmore. But why all the secrecy? Further questioning of Holt revealed nothing.

"Lloyd will give you all the dope tomorrow," was the best he could get from Holt. After a lot of talk and looking, Holt, Stark, and Rosser selected one large truck to be equipped with winch and gin-pole rigging capable, so the International Harvester man said, of handling loads up to fifteen tons, and two seven- and eight-ton tandem trucks with similar winch and gin-pole equipment.

Back in the office after lunch, Rosser made a reservation on the Frisco to Madill, Oklahoma, and called the Ardmore office to have someone meet him there on arrival Saturday morning at about 4:00 A.M. Since war restrictions on car travel were so tight, Noble employees used the train to Madill a great deal when traveling to Ardmore. Rosser spent the night in the smoking compartment since no berth was available. He was met at Madill as arranged and driven to Ardmore.

Signaling Rosser into his office, Noble inquired about the trip from Casper. Then leaning back in his office chair with his feet on the desk and pausing to nibble on the unlighted cigar he held in one hand, Noble asked, "Gene, how would you like to take four drilling rigs to the British Isles?"

Gene's knowledge of geography at the time was not very extensive. "Where the hell are the British Isles?" Rosser wanted to know.

"Well, Gene," Noble replied, "if you don't know, I doubt if I can tell you exactly where they are. But anyway that's where we are planning to go with four rigs for the purpose of drilling 100 shallow wells ranging from 2,300 to 2,500 feet in depth. Ed and I would like to have you take charge of the project and manage the drilling operation.

"The work is going to be in a war zone, and you will be working under war conditions and war restrictions. It's going to be a really tough job. The British outfit we'll be working for is going to furnish the drilling equipment which we will buy new for them in this country. We will select the type of equipment and recruit the drilling crews who will be on our pay roll as Noble Drilling Corporation's employees subject to our supervision. Fain-Porter Drilling Company will be our partner in this deal. We will be the operators and will run the show but Fain and Porter join me in wanting you to take over this job as supervisor of the drilling operations.

"If you take the job, you should get busy right away recruiting the drilling crews because that's going to be tough. Available good men are hard to find, but I want every man who agrees to go with us to be told the facts about the job. I want them to know exactly what they are getting into. They are going to be working in a war zone under severe wartime restrictions. Man power in the oil fields is getting short, but the men we talk to must know the facts about the job. We must not misrepresent the situation in any way.

"There will be a barrel of red tape connected with this recruiting business. You will be required to get deferments from the draft boards, physical examinations, birth certificates, passports, vaccinations, and to attend to all the other details that will come up connected with hiring these men."

Rosser, surprised and somewhat overawed by what Mr. Noble had told him, sat meditatively trimming his fingernails. Finally he said, "Mr. Noble, you say you want me to take over the drilling operations? I want to ask you: Am I going to run the deal or will the British have their lip in the operations?"

"Well," Noble said, "if we hadn't expected you to run the outfit, we wouldn't ask you to go with it, Gene. You are the man who will have the responsibility for the job and no one can discharge his responsibility without having the authority to do so."

"Well," Gene said, "is there anything more you can tell me about it?"

Noble responded that there was not. "This deal is a highly secret matter. It is a top-drawer war secret that cannot be discussed with anyone. If the Germans found out where the drilling site was, they would, of course, give it a hell of a good bombing! I don't know anything about the location, except it is somewhere in Great Britain. The British company we will be working for has a program that calls for the completion of at least a hundred wells within a twelve-month period. That's about all the information I can give you because it's about all the information I have, except to emphasize that you and your men will be living under war conditions just as other civilians in England are doing. The British are anxious for us to get on with the job at the earliest possible moment."

"Mr. Noble," Gene finally spoke, "let me ask you one more question. Knowing what you know about this deal, and I guess you have told me all that you do know about it, would you take the job if you were me?"

Lloyd Noble's quick response was, "I certainly would."

"All right," Gene said, without further hesitation, "I'll take it then. What's the first thing that we are supposed to do?"

"As I said," Noble commenced, "I think the first thing for you to do is to get started on a recruiting campaign. Ed Holt has convinced the British that four National 50 rigs with unitized draw works equipped with eighty-seven-foot jackknife masts working two four-man crews on twelve-hour tours will get the job done. A man by the name of C. A. P. Southwell, managing director of D'Arcy Exploration Company which is owned by the Anglo-Iranian Oil Company, Ltd., has been in New York for about two months discussing equipment and the terms of a drilling contract with Ed Holt, George Otey, and Percy Johns. He has returned to his home in England and left his lawyers in New York to button up the details. In addition to the drilling crews, you will, of course, need a good tool pusher, some truck drivers, welders, and perhaps an extra man or two.

"We will have priority over the military and should be able to secure prompt deferments from military service for the men from their local draft boards without much delay. The men will have to be recruited in the oil fields because we must have men with experience. It will be up to you," Noble emphasized, "to hire every man who goes over with you on this job, except one man."

"Who's that?" Gene wanted to know.

"His name is Don Walker," Noble informed him.

"Who the hell is Don Walker?" Gene asked.

Noble explained that Don Walker was a man who had lived in Ardmore a number of years and had a number of friends in Oklahoma. As a young man he had rubbed elbows with roughnecks at Healdton, Ragtown, Loco, and Ringling, served in World War I, attended the University of Oklahoma, and knew Frank Porter from the days they had been associated with the Wirt Franklin Oil Company. Recently, he had been employed by the Consolidated Aircraft Company, San Diego, California, as a classified war worker. "Porter has made arrangements for him to come to Oklahoma to be your assistant."

"Does he know anything about the drilling business?" Rosser wanted to know.

"Not a damn thing in the world," Noble tersely answered.

"Well," Rosser said, "he'll damn sure make a hell of a fine assistant."

"Well, Gene," Noble's blue eyes twinkled, "we hired Walker to look after you. You will find him a fine gentleman and a man you can depend on and work with. There's going to be a lot of detail here in this country as well as over there in England that he will look after. He is a good man, so don't get your 'worry hump' up about that. In fact, he will be here to go with you to help recruit the drilling crews and other men you'll need for the job. He will be ready to meet you any time and any place you say. It will be necessary to hire men who are not working now, because we must not take men from other drilling contractors who are now working."

"Well," Gene said, looking at a wall calendar, "if that's going to be our first job, you can tell this bird Walker I'll meet him at our office in Houston next Tuesday morning, November 17."

"Ed Holt will be looking after this matter and is the man you will report to. You should talk to him before doing anything," suggested Noble.

Gene said he intended to return to Tulsa and would be there Saturday night.

Sunday morning he called Holt from his Hotel Tulsa room to say he was back in town and would like to see him the first thing Monday morning. Rosser said he was to meet Walker in Houston on Tuesday. He wanted to talk with Holt about what kind of agreements were to be made with the men hired for this English thing. Their wages, method of payment, and a few other details he knew would come up, and just exactly how much these men could be told about the job. The rest of that Sunday, November 15, 1942, Gene spent writing a long letter to Bernice and reading the Sunday paper.

There was a lot of war news, but the thing he noticed most was the jump in United States oil rigs at work and the number of new locations that had been announced. Sunday night he slept well. It was the first good night's sleep he had been able to get in several days.

Monday morning he was up early, strangely excited about the new venture. Holt was already in his office when Rosser arrived. At once they got into the British deal. Holt said there was one hell of a lot of matters yet to be settled before a contract could be completed, but he thought it was a good idea for Gene and Walker to get going on rounding up the drilling crews. About noon, Noble called from Ardmore to say that Walker would not be able to meet Rosser on Tuesday, but instead Walker would come to Tulsa and meet him at Holt's office the following Tuesday, November 24. Gene welcomed the week's time before starting the recruiting. He would have a chance to go over the equipment that had been bought and discuss with Holt the equipment yet to be secured. It also gave him time to get in touch with his own draft board in Louisiana for a deferment. He had previously been classified 4-F because of an old injury to his right shoulder; getting a military deferment for himself under these circumstances should not be difficult. He had plenty to do with the Tulsa office the rest of the week, talking to Holt about the job that had been named by those in the office working on the matter as "The English Project."

Near the end of October, Frank Porter had written his old friend, Don Walker, at San Diego, that he and Lloyd Noble had taken on a deal which was now "in the hopper." He believed Don would be interested because it involved highly important work directly associated with the war and probably provided a greater opportunity for service to the war effort than the work he was now doing. He could not tell him what or where it was, but he would call him as soon as he knew a little more about the deal. Before taking the Consolidated job, Walker had talked with Porter and urged him to let him know if he should hear of more direct or more important war work than he would be doing for Consolidated. So, in view of Porter's letter, Walker was not surprised when on Sunday morning, November 8, while breakfasting around the corner from his boarding house at the cafe commonly referred to by the regulars as "the greasy spoon," he received a telephone call from his landlady that long distance was calling him. Walker hastily finished his hot cakes, margarine, and coffee; butter, sausage, and ham being out of the question. He rushed back to the boarding house to take Porter's call. Porter was indeed vague and indefinite as to the nature, character, and location of the work Walker would be expected to take on, but said he could assure Don that it was something he felt Walker would want. Definitely, it was an opportunity to perform a worthwhile war job that was more important than a routine assembly job at Consolidated Aircraft. The details, Porter said, could not be given him on the telephone, but if Walker was willing to take the job on Porter's judgment, he suggested that he give the required notice for leaving Consolidated and be in Oklahoma City as quickly as possible. Walker had been classified as a war worker and two weeks' notice of leaving the job was necessary. Walker had great respect for Porter's judgment and felt Porter would not be calling unless the job was really important to the war effort.

On Monday morning he arranged an interview with the personnel officer of his division whom he found cooperative. Consolidated would release him on one week's notice. Fortunately, Walker was able to get plane reservations out of Los Angeles on American Airlines on the nineteenth and wired Porter that he would meet him at the Fain and Porter offices in

Oklahoma City on November 20. On arrival he went directly to Porter's office where he was informed of the English project, that he would be employed as the assistant to a fellow by the name of Gene Rosser, a Noble man who would have direct charge of the operations. Don's job would be to assist Rosser generally in recruiting personnel and in the paper work connected with the job in England. The work was expected to last at least twelve months. On the job in England, Walker would assist in arranging housing for the men, obtaining food supplies, handling the pay roll and drilling reports to the Noble office, and a thousand and one other things that would arise that no one could now foresee.

Porter said that he and Walker had an appointment with Rosser and Ed Holt of the Noble Drilling Corporation and Lloyd Noble in Tulsa the following Tuesday, the 24th. The time suited Walker because he needed the weekend for a trip to Ardmore to arrange the closing of his bachelor apartment and for taking care of a few matters that needed attention before he left the country. His rush trip to Ardmore resulted in a quick rental of his apartment for twelve months and a sale of most of his furniture and household equipment. Such items as he could not sell were either given away or farmed out to friends. At any rate, Walker was back in Oklahoma City on Monday, the 23rd of November.

As planned, Porter and Walker made the late afternoon train from Oklahoma City to Tulsa and met Noble, Holt, and Rosser in Holt's office the following morning. Walker and Rosser met for the first time. Neither could know that the meeting was the beginning of a lifelong friendship and that they would be employees in responsible positions for the same corporation for many years to come. They were both briefed on the negotiations between the parties up to date and agreed on a program in which Rosser and Walker were to start at once recruiting the necessary crews. Holt estimated that not more than fifty men would be needed and possibly fewer. Noble repeated what he had said to Rosser: that it should be made clear to the men that they would be working in an active war zone under war conditions and that there were inconveniences that go with such a job, that they would be working twelve hours a day seven days a week, they would be subject to the British authorities while in England and to war restrictions as to food, clothing, fuels, and other essential items. They would be subject to British regulations as to censorship and, above all, they must be impressed with the necessity of absolute secrecy. They were not to discuss this project with anyone except their immediate families. They would be expected to execute work contracts covering a twelve-month period. Pay would be on the same basis as the English drilling crews working in the same area. Since, however, the pay scale was higher for similar services in the United States the difference between what they received in the British Isles and the total wages due for the same work in the United States would be deposited to their account in United States banks designated by them.

Holt said that the British were very anxious and impatient for the work to start, and Rosser and Walker should speed up the employment program as much as possible, but at the same time take due care that men of experience, integrity, ability, and in good physical condition should be employed where possible. The employment of men with heavy family responsibilities should be avoided. Unmarried men would be preferred.

The British were insisting that the material and equipment should be shipped from the United States not later than the end of December, earlier if Noble could get delivery and transportation arranged. It had been agreed by all parties that the Noble Corporation would have over-all supervision of the project, including the accounting procedures. Rosser would report directly to Holt, Noble's general superintendent. The

45

men hired and the equipment selected and purchased would be moved to New York for transportation to Great Britain at the times and in the manner arranged by B. R. Jackson, attorney and representative for the British company.

As Rosser and Walker left Tulsa the next morning, they could not visualize the adventures and experiences that they would share with the men with whom they would be associated in this important World War II secret project.

V.

Recruiting

Noble, Holt, Rosser, and Walker's meeting in the Tulsa office had been to consider where the best possibilities existed for enlisting personnel needed to man the British project. It had been suggested by Holt that drilling activity in the big Lowden and Salem pools of southwestern Illinois had passed its peak. Some of the rigs in Illinois, he said, were shut down awaiting removal to other areas. It appeared that experienced oil-field workers might be available in the Illinois towns of the Lowden and Salem areas. The Noble Corporation had been active in both the Illinois fields and some of the personnel at the division office at Centralia could be helpful in contacting men for the British job.

Rosser and Walker headed for Centralia to discuss the recruiting program with J. E. (Blackie) Manning, superintendent of the Illinois division. At once the recruiters began making lists of names and addresses that Manning considered possibilities. Other Noble employees, particularly Pete Hinds, a tool pusher at Saint Elmo, Illinois, added names to the list. Rosser and Walker were able to use one of the Noble cars at the Centralia office that made it possible for them to contact men living in neighboring towns and on the back roads who could not easily be reached by trains or buses. The meetings were usually small, consisting, in addition to Rosser and Walker, of five or six men who indicated an interest in foreign employment. Not more than ten were assembled at any one of the Illinois meetings.

At the close of the conferences, Rosser told the men their local draft boards would be petitioned for deferments but no one should accept the proposed employment hurriedly and without careful examination of the nature and character of the job.[1] He urged the men to discuss it with their families and return for a meeting on the following day at which time they could make known their decision as to the acceptance or rejection of the proposed employment.

Each employee, of course, would be required to obtain a passport and a British visa.[2] The employment would be in the nature of a work contract. Wages would be the present hourly scale approved by the Office of Price Administration in the United States for the classification applicable to each employee. The men would be organized in four-man crews consisting of a driller, derrick man, floor man, and an engine man or helper. They would work twelve-hour tours seven days a week. If anyone voluntarily left the employment before the end of the work period, he would pay the cost of the return trip to the United States or it would be deducted from any wages due the employee. If the employee, for any reason, should be deemed unsatisfactory by the supervisor in charge of the project and returned to the United States, cost of his return passage would be paid by the employer. As always, the prospective employees were warned that the project was a top war secret, and they should not discuss it with anyone except their immediate families.

---

[1] Occupational Bulletin 15 issued August 15, 1942, by the Selective Service System recognized the drilling industry as one essential to support of the war effort. But the bulletin did not recognize deferment for ordinary "helpers" and "roughnecks," skilled rotary floormen almost as important to efficient drilling operations as the driller himself.

[2] "Knowlton to Noble," November 25, 1942, *E* file.

Recruiting personnel for the United Kingdom job continued through December and by the end of the month, twelve former Noble and Olson Drilling Co. employees had been signed up, and nine others were rejected. Leaving Walker to help the men prepare and file application with their respective draft boards for military deferment and to assist them in obtaining the necessary information for filing application for passports, Rosser went south to contact possible prospects in the oil fields of Oklahoma, Texas, Louisiana, Mississippi, and Arkansas.

Early in January, Walker also was in Noble's offices at New Orleans to follow up and help in procuring military deferments for the men who had signed up for the English jobs. By the middle of January a total of sixty-nine men had been interviewed. Of these, forty-two had signed applications for employment. Rosser and Walker made a total of forty-four. However, deferments from the local draft boards and passports were not coming through as fast as had been expected.

Pressure for speed in getting the British project launched had been building up since Southwell returned to England early in November. Operation Torch had been activated on the morning of November 8, when Allied troops started landings on the North African coast. Literally oceans of oil were required for these troop movements, operations which further depleted British oil stocks.

The use of petroleum products in England had been reduced to a minimum with civilian use being permitted only for the most essential purposes. England's desperate need for the development of her oil fields was reflected in Jackson's messages to Ed Holt urging the departure of Walker and Rosser for England. Noble and Fain-Porter, however, were adamant in their refusal to ship out anyone until the work contract between them and the D'Arcy Company was completed. The tempo of the war and the pleadings coming via Jackson from

Southwell certainly made the urgency of the situation plain. The failure of the local draft boards to act on requests for deferment and, consequently, the hangup in issuing passports by the State Department caused Noble and Fain-Porter companies to make strong appeals to the appropriate agencies for help.

Likewise Jackson, Anglo-Iranian's representative in New York, commenced urgent appeals to the British Petroleum Mission in Washington to enlist the aid of Washington officials. They must be made to understand that launching the British oil-field project was vital to maintaining the advantage the Allies now seemed to have. The need of British civilians for an additional quantity of petroleum was desperate. England was actually "scraping the bottom of the barrel" with storage-tank bottoms being dug out and carted to refineries.

Cartwright Reid, Anglo-Iranian's representative in Washington, and Frank Porter, who was spending considerable time in Washington as president of the Mid-Continent Oil and Gas Association, lost no time in conferring with Knowlton and Mr. James Terry Duce, head of the foreign section of PAW. Knowlton and Duce went into action with a barrage of letters, telegrams, and telephone calls to state directors of the Selective Service System. General Lewis B. Hershey, military head of Selective Service, was contacted, in cases where the men were refused deferments or where the local boards had failed to take action on pending applications.

The impact of the Washington officials upon the state directors of Selective Service and the local boards was dramatic.[3] The log jam appeared to have been broken. Even orders of men who had been processed by their local boards for induction into military service, but had not yet been inducted,

[3] "James Terry Duce to Edgar Holt," November 28, 1942, *E* file. This file contains letters from the directors of Selective Service in Illinois, Louisiana, and Oklahoma that local boards had been requested to grant deferments, as required, for the men selected for the overseas project.

were recalled and the applicants given deferments and permission to leave the United States for the maximum period of six months. At the end of the six-month period, draft authorities advised, it would be routine but necessary to obtain extension of the deferment period.

By the end of January practically all the men had received their deferments and permission to leave the country. Two had been inducted into the service, and therefore were no longer under the jurisdiction of the local draft boards. Three others had changed their minds and declined the employment. These men were hurriedly replaced and now forty-two men were assured for the project. Knowlton had previously arranged with the State Department that applications for passports covering the drilling crews were to be sent directly to the attention of the chief of the passport division, Department of State.

Rosser and Walker had finished with details in the field, and Rosser made a quick trip to Alvin to say good-by to his wife and child. Returning to Tulsa, January 23, to pick up loose ends, he left Tulsa the same day for New York, arriving there Monday morning, January 25. Rosser contacted Jackson and his staff at once and commenced a check with the National Supply Company men headquartered in New York as to the drilling equipment that had been purchased and received in New York for shipment to Britain. Since Jackson had pending with the British Petroleum Mission in Washington and the military authorities an application for immediate air passage to Britain for Rosser and Walker, Rosser knew that he had only a few days for checking the equipment. He was disappointed to find that National Supply had been unable to secure and deliver the four Hesselman–Waukesha engines, which meant two rigs were without operating power. National was also short on the delivery of drill pipe. He was advised by National Supply that no hand tools such as axes, hammers, wrenches, crowbars, and the like could be found. The failure of National Supply was actually attributable to shortages and not to lack of any authority to furnish the equipment. Duce informed both Holt and Jackson that the British project had been assigned an AAA priority by the War Production Board, the agency which allocated equipment in short supply.

Rosser was amazed to find that only one gross of canvas gloves had been purchased, and he disdainfully declared, "Hells bells, Mister, one gross of gloves wouldn't last our drilling crews a week." National Supply was asked to secure immediately one hundred gross of gloves for the project. As ridiculous as it may seem, this was one of the hardest jobs assigned to National Supply, but with the assistance of Jackson's office and Noble's Tulsa office, the men were finally supplied with the 14,400 pairs. Manufacturers and local oil well supply stores were enlisted in the effort.

Rosser was considerably worried about the hand tool situation. One of the clerks in National Supply said that he had heard of a retail dealer somewhere in Brooklyn who had a supply of hand tools. This hint was all that Rosser needed. A subway trip to Brooklyn and a taxi cruise up and down Brooklyn streets for a full day produced no results. On the second day of reconnaissance Rosser was surprised and overjoyed to pass a double-front hardware store that appeared from the outside to be productive. On entering the place, he was awed by the great quantities of small tools stacked on counters, shelves, and even hanging from the ceiling. The proprietor was ready to sell anything he had, but his business was cash! "You give me the cash, and I give you the tools," was his explanation to Rosser. "No checks or charge accounts." Rosser's inquiry as to priorities or price controls meant nothing to the owner. His response was, "I need no priorities, and I sell for what these tools cost me plus my profit. I don't know nothing about price controls, and I ain't fixin' to learn."

Rosser realized that Lady Luck had smiled upon him and he was certainly not going to pass up the bonanza he had stumbled onto. Rushing back to the National Supply office in New York, he had little trouble arranging for a supply of cash. Back in Brooklyn, he began negotiations with the hardware man immediately. Axes, hammers, wrenches, chisels, saws, crowbars, and other tools of various sizes, types, and makes were soon piled high; Rosser triumphantly laid down the cash to cover the total price. Taking no chances on a slip-up in delivery, he called two taxis. Loading both of them to the ceiling, Rosser climbed into the front seat of one of the cars and led the way back to the National Supply Company's store in New York, where the amazed personnel received the tools for packing preparatory for shipping.

To this day Rosser and some of the National Supply men are still wondering how it was possible for either the hardware owner or customers to do business in open violation of wartime rules and regulations. But since the statute of limitations has long since run out against the crime, and since the hand of fate seems to have directed Rosser to the right place at the right time, and since it was all for the war effort of Great Britain for survival, no serious pangs of conscience bothered either the Americans or the British directly or indirectly involved. The incident still brings a warm smile to those who knew about it at the time.

It had become apparent in early January that National Supply was going to have trouble supplying some of the equipment of the four drilling rigs and particularly the Hesselman–Waukesha diesel engines and the Caterpillar tractor equipped with grading blade. On January 15, Jackson wrote Holt saying he had just returned from a four-day visit to Washington. The AAA priority issued by the War Production Board for the project should insure delivery of the engines within the next eight to ten weeks. However, to speed up delivery, he had asked the army which was taking the entire output of the seventy-five Waukesha engines per month, if eight engines necessary to power the four rigs could be released on the condition they would be later replaced by D'Arcy.

He was hopeful that the army could be persuaded to give this much assistance. Transportation, he advised, had been arranged for the equipment. As insurance against the submarine menace, four vessels would carry a cargo of one drilling unit each. In the same communication, Jackson advised that transportation for Rosser and Walker would be available during the first two weeks of February, and again urged they be sent ahead as quickly as possible. Holt assured Jackson that Noble and Fain-Porter were ready to release them as soon as the contract could be concluded. Jackson again wrote and called Holt on January 22, saying he had suggested to Otey that a meeting should be arranged in New York about February 1 to conclude the contract.

Holt, Otey, and Walker joined Rosser at the Lexington Hotel on Tuesday, February 2. Negotiations of the parties continued through Friday of that week. Saturday morning, the 6th, Ed Holt awoke early as was his custom, and when Ed said early, he meant not later than 5:00 A.M. The closing of the contract and getting back to a mountain of wartime problems awaiting his attention in Tulsa weighed heavily on his mind. As he reviewed the points of the contract that had been discussed, and reread for the twentieth time the memo of Percy Johns enumerating item by item the matters that must be settled in the final contract, Holt came to the conclusion that a great deal of talk was going on about many nonessential points. To him it appeared that the method and manner of covering the drilling crews and the supplemental personnel on the United Kingdom project with adequate insurance about which both Noble and Porter had been particularly concerned was really the only important issue left to be decided. With this in mind,

Holt hurriedly bathed, shaved, and dressed. Before 6:00 A.M. he was rapping on the respective bedroom doors of the Noble group demanding: "You fellas get dressed and come into my room as fast as you can." Within a few minutes, Otey, Rosser, and Walker were in Holt's room wanting to know the urgency of this predawn gathering. Holt swiftly reviewed the points in Johns' memo yet to be settled. Actually, he felt nothing was of serious importance except the insurance clause. Holt's patience was short on this Saturday morning, and as he talked, it became shorter. When someone asked, "Well, what the hell can we do about it?" this was enough to light Holt's short fuse and send him into action.

Leaping to his feet and pounding the night table by his bed, he exploded, "We are going to Jackson's office as soon as we get some breakfast. We are going to be on his doorstep when he gets there. I am going to tell Brother Jackson that this whole damn deal is off unless we can button up the contract today, and if we don't get the son-of-a-bitch settled today, I'm packing my dirty shirt and getting out of this damn town. I don't like it anyway, and I've got a lot of other things to do. This is not the only thing we've got on our slate."

The group ate breakfast in silence and went straight to 620 Fifth Avenue. Jackson was coming into the building as Holt and his companions piled out of a taxi. As soon as everyone was seated in Jackson's office, Holt reviewed points not yet settled in the proposed contract. Most of these related to reporting and accounting procedures, and he emphasized that all of these unsettled details could and must be agreed upon today. Otherwise, he must leave to get back on the job to attend to a number of pressing matters awaiting him.

Jackson, sensing the tenseness of the situation, listened carefully. When Holt had finished, he swept away his British reserve and traditional caution. Point by point, the pending items were either agreed upon, or such procedures left to be worked out between Jackson's office and Johns of the Noble office. Their decision would be acceptable to the parties. Finally, only the insurance matter remained. The insurance brokers had previously proposed that the insurance companies be requested to issue a policy or policies that would give the Noble and Fain-Porter employees the same coverage as that provided by the workmen's compensation laws of Oklahoma. Otey said he felt such an arrangement was fair and reasonable. Holt and Jackson agreed.

With this matter concluded, Jackson closed the big notebook that lay on his desk before him with a bang. He rose, walked around his desk, and shook hands with everyone, announcing that so far as he was concerned the contract had been concluded. All had complete confidence that a meeting of the minds had been had, and no one doubted that a contract acceptable to the parties would now be prepared at once. This was not to be confirmed by the execution of a formal contract until many days later, pending the procedures to be worked out between the Noble and Jackson offices, together with the appropriate forms concerning reports, billings, payments of salaries, wages, expense accounts, settlement of insurance claims, drilling and completion records, and numerous other details. Although actually executed at a later date, the lawyers, remembering the swift conclusion of the agreement between parties on that February morning, took pleasure in dating the contract: "As of February 6, 1943." Holt and Otey were able to leave New York for Tulsa on Sunday, the following day.

The principal barrier to the departure of Rosser and Walker had now been removed. Jackson confirmed with the Anglo-Iranian's representative in Washington that a travel priority for Rosser and Walker on the American Airlines Yankee Clipper out of Montreal to London was being forwarded that day effective for the night flight of February 20. Jackson, knowing the uncertainty of securing air travel priorities for any particular flight, had taken the precaution of also making reservations for Rosser and Walker for surface travel.

This was for passage February 10 on His Majesty's ship, the "Stirling Castle," out of a South African port. At noon on February 8 American Airlines called Jackson's office to advise it would be impossible to honor the February 20 travel priority of Rosser and Walker on the Clipper flight from Montreal to London on that date, because all space had been pre-empted by the army for certain people traveling pursuant to military orders.

Although Rosser and Walker certainly did not relish surface travel in view of the intensified submarine wolfpack attacks on shipping in the Atlantic, they realized, however, they had a war job to do and accepted the news from Jackson's office without complaint. Previously they had been armed with the appropriate travel priorities, berthing tickets, passports, permission to leave the United States, proof of vaccinations, health certificates, censorship clearance from the Board of Economic Warfare—Technical Data License Division, and letters from Jackson containing directions concerning their arrival in England, and also clearances that would be required by both British and American officials. Contact was to be made with Cook's Travel Agency, Ltd., on landing. Walker had spent considerable time in Jackson's office being briefed on these matters, and now felt that he was familiar with the intricacies of wartime travel.

Rosser now remembers that Walker was a walking encyclopedia concerning the complexities of travel at that time but could not even talk to himself because of the strict supersecrecy that had been imposed on both of them. They had been warned that the New York waterfront was alive with German agents and their confederates. The bulletin that came with their travel priority requested in large bold type that for their own safety and protection of the lives of others traveling with them, that they were not to discuss their travel plans with any person.

They were to board the "Stirling Castle" before 10:00 P.M.[4] Shortly before that hour, Rosser and Walker presented themselves at the ship's office on Pier 80 at the foot of 36th Street, and after their credentials were examined, they were given boarding passes. The realization of war came with forceful impact when they walked up the gangplank bristling with armed guards in the uniform of the United States Marines. They were warned by one of the military police that once aboard they would not be permitted to leave the vessel until it docked at destination. What destination Rosser and Walker had no idea. They knew they were bound for some place in the British Isles, but where the "Stirling Castle" would be docked was known only by the ship's captain and mate, and even such destination could be changed by wireless code at any time.

In the meantime, Holt found the shoe on the other foot—it was now Noble and Fain-Porter's time to urge speed. The boys who had been signed up for the project were getting restless. Twelve-hour tours were being adopted throughout the oil fields. Manpower was scarce. The boys were asking, "Why leave when you can stay home and get twelve-hour tours?"

On January 22, Jackson advised that transportation had been assured for the men by surface travel, in two groups. The first group would go about the first week in March and the second about a week later. Because of the uncertainty as to dates, he requested Holt to send the first contingent of sixteen members of four drilling crews, plus truck drivers, swampers (helpers), one tool pusher, and chief mechanic, making a total of twenty-four men, on to New York. Accordingly, Holt sent out notices to this group asking them to be in Tulsa on

[4] Rosser kept a brief diary relative to his English venture. His entry of February 10, 1943: "Sailed on the big water Atlantic from New York. Expecting to be seasick before we clear the harbor. The name of the boat is 'Stirling Castle.'"

February 23. As of that day, they would go on Noble's payroll. On the following day, February 24, the first contingent of twenty-four men left for New York under the supervision of Holt's assistant, Howard (Red) McCarty. On the morning of February 26, the first group was registered in at the Hotel Victoria on 7th Avenue at 51st Street, New York City. The experience, initiative, and leadership qualities of the "boys" had been analyzed by Rosser who made recommendations as to the classification of each. These had been reviewed by Holt. Accordingly, the men had been organized into crews with classifications as follows:

Noble Drilling Corporation
and
Fain-Porter Drilling Company

PERSONNEL—UNITED KINGDOM PROJECT

Driller	Johnson Wilson Nickle
Derrickman	Gerry Ernest Griffin
Helper	Joe Townsen Webster
Motorman	Ester Glenny Gates
Driller	Virgil Lee Latham
Derrickman	Albert Francis Webster, Jr.
Helper	Clinton Francis Johnson
Driller	Allen Johnston May
Helper	Ray Franklin Miller
Helper	Aaron Levi Long
Motorman	Woody Wayne Walden
Driller	Everett Harrison Hemphill
Derrickman	Edward Howard Boucher
Helper	Albert Alexander Morton
Motorman	John Henry McIlwain, Jr.
Driller	Phillip Earl Albritton
Derrickman	Herman Douthit
Helper	Clement Mathais Riedinger
Motorman	Joseph Lafayette Waits

Driller	Horace Glendon Hobbs
Derrickman	Elray R. Davis
Helper	James Edward Harding
Motorman	George Garrett DeArman
Driller	Lewis Vernon Dugger
Derrickman	William Merl Burnett
Helper	Eston Everett Edens
Motorman	Raymond Allison Hileman
Driller	Loyal Melvin Oaks
Derrickman	Dewey Aycock
Helper	George Christie Watson
Motorman	Joe Simmons Barker
Truck Driver—Class A	Loran Penn Robinson
" " — " A	Kenneth John Johnston
" " — " B	Clarence Urban Sikes
" " — " B	Fred Emery Moss
Truck Helper—Class A	Carl Gustaf Norberg
" " — " A	De Talt Havely
" " — " B	Robert Milton Christie
" " — " B	Curtis Elgar Olvey
Supervisor	Eugene Preston Rosser
Toolpusher	Gordon Oscar Sams
Chief Mechanic	Luther Berry McGill
Ass't. Supervisor	Donald Edward Walker

That there would be changes made in the crews, as time went on, was certain. But for the present this organization of the crews would stand. Loyalty, cooperation, and efficiency were the prime requisites of a good crew. For that reason, Rosser and Walker took care to name the men who would be acceptable to the driller.

On March 6 Jackson telephoned Holt telling him that after advising with the British Petroleum representative in Washington and the Army staff, the British Merchant Shipping Mission, and the air authorities, and acting on their advice, the entire group would be dispatched by a very special surface plan by March 15. In view of this latest arrangement, Holt notified the second group to be in Tulsa on the morning of March 8. The following day, March 9, eighteen men including drilling crews 5, 6, 7, and 8 and two extras, entrained for New York arriving there Thursday, March 11. A total of forty-two men were now in New York ready for embarkation to somewhere in the British Isles.

Gordon Sams, as tool pusher, at this point took over the entire personnel together with the mass of records required for wartime travel, including passports, all for delivery to Walker on arrival in the British Isles.[5] Like Rosser and Walker, Sams was furnished with full instructions for contacting Cooks Travel Agency on landing and the Anglo-Iranian Company offices in London if he found it necessary or desirable.

The first group had their time in New York. From fragmentary bits of available information, they had made the most of it, such as taking over a Chinese restaurant in Chinatown and turning handsprings through the amazed onlookers on a subway platform. One of the "boys" had made the entire length of the loading area in a fine display of backward flips. The plate-glass mirror covering one wall of a private dining room set aside for the "boys" at the Victoria Hotel was somehow broken. But so far as is known, not a man had said a word about the United Kingdom project. No drink was strong enough to loosen their tongues sufficiently to mention this top war secret project in which they were now involved.[6] "Red" McCarty, a former football star at Oklahoma University and an all-American selection by several of the national magazines and news services, who had "ridden herd" on the group,

[5] "B. P. Jackson to Edgar Holt," Western Union message, March 5, 1943, *E* file.

[6] Reminiscences of Lewis Dugger and others who were eyewitnesses to and participants in the English project.

is said to have collapsed into bed for two days after the "boys" sailed.

The second contingent was sorely disappointed. They did not have an opportunity to "do New York" because time and war waited for no man. Shortly before midnight of Friday, March 12, the "soldiers of oil" were marched aboard H.M.S. *Queen Elizabeth,* which had been converted to a troop carrier and was docked at Pier 90 at the foot of West 52nd Street.

VI.

"This England"

With the main essentials of the contract between Anglo-Iranian and the Noble and Fain-Porter companies completed, Southwell had felt his mission accomplished. Remaining details that involved travel priorities and transportation could be left to B. R. Jackson, the Anglo-Iranian attorney and representative in New York, and to his office staff. Agreement as to forms, methods, and manner of operational reports to be furnished Noble and Fain-Porter companies, Southwell felt, was a mere matter of detail that had little bearing on the fundamentals of the contract. It had previously been agreed that Don Walker, now designated as assistant supervisor, would be in charge of the administrative work.[1]

Paramount in Southwell's mind was getting the equipment and man power on location and developing swiftly the additional production from the recently discovered British oil fields. Also, he was extremely anxious to drill a number of outside test wells to deeper horizons with the hope that new production might be discovered. Anglo-Iranian geologists and production men were enthusiastic about this phase of the British oil picture, as geophysical data collected in the areas appeared favorable. Despite Southwell's desire to get some of

---

[1] See "George Otey to B. R. Jackson," November 16, 1942, and Jackson's reply on November 19, *E* file.

63

the outside wells started in order to test new and deeper geological formations, he realized that Britain's critical need for oil was such that the first efforts must be directed toward completion of additional wells in the already proven areas.

The problem of compelling importance at the moment was finding and arranging necessary housing for the forty-four oil-field workers soon to be arriving in England. Certainly, the project should be well under way by the end of the year. Southwell was thinking of the many tasks that had to be done as he came down the steps of the Yankee Clipper that had touched down soon after daybreak at London's Croydon Airport. As Southwell pushed his way through the jam-packed lobby of the airport, he came to a sudden stop, his attention caught by big black headlines in the morning papers at the nearby newsstand.

Southwell needed to get a tight grip on his emotions and calm himself with a couple of good deep breaths before he could comprehend fully the bold headlines. Slowly he began to realize the significance of what he was seeing. The noisy chatter of the crowds that swirled around him made concentration difficult. A calmer look at the morning papers told him that United States and British troops had landed in North Africa.

At 3:00 A.M. of this November 8, 1942, more than 800 war vessels and transports moving convoys off the coast of North Africa commenced disembarking 107,000 British and American troops simultaneously at Casablanca, in Morocco, on the African Atlantic coast; and at Oran and Algiers on the Mediterranean coast of Algeria.[2] Southwell hurriedly devoured the front-page bulletins despite the drizzle of rain that trickled down the back of his neck inside his collar, then managed to

---

[2] C. L. Sulzberger, *World War II*, 136–39. Winston S. Churchill, *The Hinge of Fate*, 606–20.

crowd into a taxi for transportation to the Anglo-Iranian offices.

Directors and key officials of that parent company were summoned to a meeting to hear Southwell's report on the two months he had spent working out the complex details of a drilling agreement with the American companies to furnish drilling crews and to direct drilling operations for completing 100 wells at the locations to be selected by D'Arcy over a period of twelve months. If well completions could be carried out as rapidly as Mr. Ed Holt of the Noble Corporation had insisted, he assured the group, the 100-well program into the 2,500-foot producing horizon could be easily completed within the contract time of one year. Despite Southwell's enthusiasm and the accomplishments he now visualized in the fields, it was difficult for the men of the parent company, who had been conditioned since 1939 to the slow pace of well completions, to really believe what they were hearing.

Southwell went into considerable detail about the new and latest equipment being purchased, equipment which he had been privileged to see at work in the Lowden and Salem fields of the Illinois basin.[3] Operation of the drilling rigs would require a minimum of labor personnel for a number of reasons; the fact that the drilling rigs were assembled as a unit gave them maximum mobility of movement from one location to another. The jackknife drilling mast would avoid the time now consumed in building and tearing down the old conventional wooden derricks. A maximum of speed with a minimum of personnel, Southwell said, was insured by the friction hoisting clutch and the excellent brake on the drum. A small Shaffer spinning cat-head for spinning up the drill pipe, although not standard equipment, was incorporated, and with a spinning

3 "Southwell to Noble," October 30, 1942, in E file relating details on the equipment and the proposed contract obviously brought to the attention of the directors of Anglo-Iranian Company.

chain resulted in great speed when making up the drilling string. The rigs, he told the Anglo-Iranian officials, would be powered either by two Hesselman diesel Waukesha 169 h.p. engines, or by two 150 h.p. electric motors.

The slush pump, he explained, was set behind the motors and driven by the Vee belts. The rotary speed could be regulated by merely changing the speeds on the jack-shaft by using the clutch lever next to the driller. Three speeds would be available.

Southwell planned one-thousand-gallon fuel tanks for each drilling rig operated with Diesel engines. It was calculated that this much capacity would be sufficient fuel to complete one well for production without the necessity of moving additional fuel to the location. Tanks, he said, would be cylindrical and mounted on skids in a horizontal position. Southwell pointed out that transportation was the key to rapid moves from one location to another in America and was achieved by using specially designed transport for which he had made arrangements—one fifteen-ton winch truck, one seven and one-eighth-ton winch truck, and two five-ton winch trucks, all rigged with gin poles and lines, so that the trucks could perform a self-loading operation. Specialists were required as operators.

The new equipment included one Caterpillar D.6 bulldozer with grading blade for excavating mud pits at the drill site. A quantity of Hughes and Reed rock bits had been purchased. One string each of three-and-one-half-inch and four-and-one-half-inch drill pipe would be used. Additional drill pipe would, perhaps, be needed before the program was completed. Southwell was happy that the equipment contracted for was capable of drilling test wells to a depth of 7,500 feet, and the company directors agreed this was of great importance in order to determine Great Britain's oil-producing potential at the earliest possible date.

After his report on what he felt had been a highly successful arrangement in the United States, Southwell took the late afternoon train for Grantham. It was good to be home once again, despite the hardships and dark shadows of war that clouded the bright sunshine of the autumn days.

Southwell could not refrain from an audible chuckle brought on by the realization that in all of the talks, conferences, meetings, and negotiations with the British and American authorities, including Noble and Fain-Porter companies, he had been able to avoid disclosing that the oil-producing areas of Great Britain actually lay within the confines of the great legendary Sherwood Forest of Nottinghamshire, an area so rich in the history of early England. The fact that more than fifty producing oil wells had been completed with a current productive capacity of nearly one-thousand U.S. barrels per day was sufficient to satisfy all who questioned him as to location of the producing fields; the mere fact that oil production existed in substantial amounts in Great Britain not only satisfied the curious, but likewise amazed them.

On the following morning, Southwell again briefly reviewed, for the benefit of his own staff, the arrangements made in the United States for men and equipment and took occasion to emphasize the growing importance of oil to the war effort and the importance of the British oil production in particular. With the United States facing the responsibility of fueling a land and naval war against the Japanese throughout the broad expanses of the Pacific, oil had suddenly assumed the proportions of victory itself for those who had access to it, and by the same token, the defeat of those who did not have it.

The helping hand of America, the world's greatest industrial giant, Southwell felt, had been extended to them. Britain's partnership with America was indeed an inspiration to the officers and directors of the Anglo-Iranian Company, as it was to the Secretary of Petroleum and other British officials. The men and women of the D'Arcy Company felt a certain thrill at being a part of this important war work. Oil men of both Britain and America now understood that oil was the answer to the Allied offensive on the many battle fronts of the world.

So it was with this sense of responsibility that Southwell commenced a tour of the Eakring-Duke's Wood area in the neighborhood of the drilling operations in order to find appropriate housing facilities where the forty-four men who were to arrive soon might be billeted. The thoughts for housing by Southwell and his associates in the area had included:[4] Kirklington Hall, a thirty-two room English Manor House adjacent to the Duke's Wood field; The Saracens Head Hotel in the town of Southwell; The Clinton Arms Hotel at Newark; and a building in the village of Ollerton, formerly used as a dormitory for a girls' preparatory school, which it was thought might be remodeled for use of the American roughnecks. There was Kelham Hall, in the village of Kelham, a monastery for the monks of the Society of the Sacred Mission, an order of the Anglican church. Parts of this huge building had been used for various military purposes since the commencement of the war in 1939.

Serious consideration had been given to the possibility of building barracks similar to those used at Great Britain's many army bases, but in view of the shortage of building material and manpower, the idea of building such housing was discarded.

Late autumn in the English Midlands is one of Britain's most charming interludes. The variety of red, bronze, and russets; the deep purple and rich browns with sprinklings of yellow and gold of the great oaks, yews, and copper beeches extending from the village of Ollerton across Nottinghamshire, east and west to the southern border and covering approxi-

---

4 Fred T. Jones, *Newark-on-Trent*, 33; Father Gregory, *Kelham*, 15–17; also, *In the Shade of Sherwood Forest*, 4, 5, and *A Concise Guide to Southwell and the Rural Districts*, 6–7.

mately 60 per cent of the area of Nottinghamshire, constituted the remaining remnants of the southern end of the once heavily wooded Sherwood Forest.

Much of the forest had now been sacrificed to the villages and pastures of the area. As Southwell drove the roads linking Newark-on-Trent, Nottingham, Kelham village, and the town of Southwell, he was conscious of the quiet beauty of the rolling hills of the countryside which may have spoken to the geologists in a language of science indicating surface evidence of possible oil-bearing structures. But on this autumn day of 1942, to C. A. P. Southwell they were simply gentle pastures.

The red brick Georgian homes, which had served the country squires and the gentry as manor houses for generations, settled into the landscape as jewels set in a bed of velvet green. Here and there the idyllic picture was brought to life by the movement of brown Swiss and Ayrshire cattle, or the slow meanderings of a flock of white sheep nibbling its way across the pasture.

The warm beauty of the Midlands countryside made the war seem far away—a happening with which its residents seemed to be disassociated. A casual observer would think of it as only the stage upon which the people of legend and history had played important roles. Robin Hood and his Merrie Men, the Sheriff of Nottingham, Little John, Maid Marion, and all the rest became very real as Southwell drove the roads under nature's cathedral arches, formed by the curving branches of the autumn-dressed trees.

To a historian, Cromwell's and his Parliamentarian followers' presence could be felt as they galloped over the countryside seeking combat with King Charles I and his Royalists. As more tangible evidence of their presence stood the remains of Newark Castle that had been commenced on the banks of the river Trent about the year 1123 by Alexander De Blois, the Norman Bishop of Lincoln. By the order of Cromwell, the castle had been destroyed by his followers. King John is said to have died there in October, 1216. On one of the walls of the Saracens Head Inn was a marker that identified the very room which Charles I is said to have occupied on several occasions, while the conflict between King Charles's Royalists and Cromwell's Parliamentarians raged. In later years, Mr. W. E. Gladstone had occupied a suite in Newark's Clinton Arms Hotel while he campaigned for and won a Conservative seat in Commons at the early age of twenty-four.

The ancient and history-drenched minster of Southwell with its spacious, autumn-blooming rose gardens, made a heady contribution to Southwell's reverie of the times of long ago when suddenly, the roar of reconnaissance planes overhead brought him back to reality. Quickly, he reviewed the places that might be appropriate for housing the American workers. He made a mental note of his resolve not to permit the destruction of, or damage to, the beauty of the Midlands one whit more than the urgencies of the war required. The idea of the war trespassing upon this autumn beauty, where so much history and legend of the British people abounded, made Southwell's blood run fast with renewed anger and hate of the war lords in Berlin. He was happy that he had insisted that the performance on the contract to augment the development of the oil fields here in the Midlands would be carried out with due regard to saving this bit of English heritage. The drilling and oil-producing equipment that was now being used by the D'Arcy Company had been painted grass green to camouflage its location, nestled under the arms of the great oaks, yews, and chestnuts of Sherwood Forest. They merged well with the pastorial beauty and surrounding countryside.

VII.

Safe Arrival

"Walker," moaned Rosser, as he stumbled through the state-room door mopping his face with a cold towel, "I am a sick son-of-a-bitch—what day is it anyway? How long do you suppose we are going to be on this damn tug?"

Walker could not resist the opportunity to twit Rosser a bit about his seasickness. "We're only two days out; it's way too soon to be seasick. Besides, the old Atlantic has been as smooth as a millpond since we left New York."

"Smooth, hell," Rosser commented, "this tub's got more ups and downs in her than a Cheyenne rodeo bull!"

"Gene," consoled Walker, "you are probably paying for what you did to the Queen's mailman."

Walker's reference to the Queen's mailman was brought on by Rosser's activity when they had come aboard the Stirling Castle and found the stateroom unoccupied.[1] Their stateroom was of modest proportions and had been converted from its original two lower berths to a room for four occupants. The crude expansion had been accomplished by adding an upper bunk to each of the lower berths. Disregarding the number on his berthing card indicating an upper bunk, Rosser had

[1] Entries in Rosser's diary from Wednesday, February 11, 1943, through February 20, 1943, are principal sources on the ocean voyage. They are supplemented by interviews of the author in 1968–69 with Gene Rosser and Don Walker.

chosen one of the lower berths. A large handbag, overcoat, heavy gloves, and a funny-looking elongated bag that Walker identified as a diplomatic pouch were all piled on the berth selected by Rosser. On the side of the pouch appeared an embossed crest of Her Majesty, the Queen of Holland. Rosser was not impressed. Unceremoniously, he gathered up the articles and pitched them onto the upper bunk; made ready for bed; and crawled into the lower berth, urging Walker to do the same with respect to the other lower berth with the advice that if anybody took issue to just let him know.

Rosser and Walker could feel the throb of the ship's engines and knew the vessel was in motion. The missing berth mates came into the stateroom. Finding both lowers occupied, they took the uppers without comment.

On the following morning, Rosser and Walker learned that the man with the diplomatic pouch was a courier in the service of Holland's Queen Wilhelmina, who had found asylum in London. The Dutch gentleman, Dr. Von Kelleher, explained for the benefit of Rosser and Walker's curiosity that the pouch was empty. They assumed the state documents, if any, intended for the Queen were safe in the purser's strong box or were carried on the person of the courier. Thus it was that Rosser had referred to Dr. Von Kelleher as the Queen's mailman.

The fourth member of the stateroom introduced himself as Jerry Harewood, who volunteered the information that he lived in Ft. Worth, Texas, and was a geologist. Beyond mutual introductions, the parties disclosed no further information concerning their business or destination. Rosser and Walker speculated that Harewood was undoubtedly, like themselves, on a war mission concerned with oil but did not feel disposed to elaborate.

On Saturday, February 13, Rosser noted in his diary: "Still seasick and puking; wish to hell they would let us open these portholes at night and get some fresh air in here!"

Sunday, February 14, Rosser was able to write: "Still woozy, but am able to take a little nourishment today. 'Mister' Walker, however, is ailing today; Walker insisted that his nausea was the result of the strong black tea, heavily laced with condensed milk and brown sugar that was served at morning and afternoon tea."

On Monday, the 15th, Rosser noted, "Good sailing," and on Tuesday, the 16th: "Boating still good."

The afternoon radio broadcast on Saturday, February 20, brought shocking news. The Yankee Clipper, on which Rosser and Walker had originally been booked for air passage from Montreal to London, had crashed in Portugal, killing fifteen passengers and seriously injuring ten others, including the popular singing star, Jane Froman.

Rosser's reaction to the news was a hearty slap on Walker's back with the comment, "Ole buddy, we don't have to be scared of anything now. 'The Big Man' upstairs has got His arm around us now! You can't ever tell me that He didn't have a hand in bumping us off that plane."

The Noble and Fain-Porter personnel, and the Rosser and Walker families, soon had their first taste of war anxiety. A German broadcast had been picked up on Wednesday evening, February 17, announcing that a British passenger-cargo vessel out of New York, bound for Europe and carrying a group of oil technicians had been sent to the bottom. *Time* magazine was to report the attack in its issue of March 1, 1943. Actually, two ships were lost instead of the one claimed by the enemy. *Time* magazine cryptically remarked:[2]

BURY THEM AT SEA: Two United States passenger-cargo ships, cramfull of servicemen and civilians on war missions, started eastward across the Atlantic in early February. Somewhere at sea, U-boats, probably using wolf-pack tactics, picked them up, kept snapping at their heels. By night, a torpedo sank one of the ships;

2 *Time Magazine Capsule 1943*, 74.

four days later, the other was sunk. Each ship went down in less than thirty minutes.

This week Washington, announcing the sinkings, also announced the death toll: more than 850. The horror came home with Signalman Robert Weikart, whose ship was the first to reach the spot where one of the torpedoed vessels went down.

Said Weikart:

We saw hundreds of bodies in the water and lifeboats full of men swirled about us. It took me a while to figure out why we did not stop to pick any of them up—they were frozen to death at the oars of their lifeboats. I saw the sea dotted with bobbing heads in life jackets. I started counting, but realized there were hundreds so I gave up. We left them there—that's the best thing. All sailors want to be buried at sea anyway.

American headlines and radio broadcasts had quoted the German announcement of the sinking. Reference to oil technicians struck horror to the mind and heart of Bernice Rosser. Those in Noble and Fain-Porter companies who knew the men were crossing felt sick and weak in the knees. Rosser and Walker were not to know until long afterwards that the United States news stories and the radio newscasts had carried the German announcement. Even if they had known, there was nothing they could have done, since European-bound ships were prohibited from using the radio transmitter except in cases of actual attack. Those at home were not relieved of their awful anxiety until the Anglo-Iranian Oil Company on February 24 cabled the news of Rosser's and Walker's safe arrival in England.[3]

Nor did Rosser and Walker know that the *Stirling Castle* had been chased by an enemy submarine causing her to alter her course some three hundred fifty miles off the regular shipping lanes to avoid a German submarine wolf pack, one of many prowling the North Atlantic. Rosser and Walker were

---

[3] "Lloyd Noble to Mrs. Eugene Rosser," February 24, 1943, Western Union message, *E* file.

only bored by the lengthening days of the rolling, tossing ship's journey.

Rosser proposed a little friendly poker game. Von Kelleher, in broken English, explained he did not know the game. Rosser replied that was fine—excellent, and added, "We'll teach you the greatest of all American sports." This he did to the enrichment of himself by some thirty dollars, to which Dr. Von Kelleher made a substantial contribution.

February on the North Atlantic was not a pleasant or easy time. The winter gales made it necessary to keep the portholes closed tightly, and regulations kept them blacked out at night with heavy curtains that the passengers were required to keep drawn from 4:00 P.M. to 8:00 A.M. Despite the chill outside, many of the passengers took frequent walks around the deck to be relieved of the smelly and crowded passageways below. The *Stirling Castle*, Rosser and Walker learned, had sailed out of Bombay before Christmas, 1942, carrying a group of homeward-bound British soldiers and civil servants, wives, women, and children, together with their personal belongings piled high along the sides of the ship's passageways.

British military and civilians in foreign service after ten years were entitled to return home for reassignment or retirement. At Rangoon, Burma, a detail of British army nurses trying to reach London had been picked up. The long voyage had made weary travelers of the Britishers from the Far East.

At last, on Friday, February 19, the exciting news spread through the ship that land had been sighted. On Saturday, February 20, shortly after lunch time, Rosser and Walker marched down the gangplank onto British soil and for the first time learned that they had been headed for and were now in the west coast port of Liverpool, England. Those debarking at Liverpool were herded into the customs area of the Port Authority. Rosser's diary records that they had a "hell of a time" getting through customs. He also noted that he had invited a young United States Army lieutenant and several shipboard passenger friends he and Walker had accumulated on the voyage to the Adelphi Hotel in Liverpool for cocktails and dinner. The American liaison officer had been able to secure a room for them while they were busy clearing customs. Although smoke and dark rain clouds hung low over Liverpool's waterfront, it was surely good, they felt, to have one's feet on solid ground again.

Rosser and Walker noticed, for the first time, that the *Stirling Castle* carried large white crosses painted on either side, proclaiming she was an unarmed nonbelligerent vessel and carrying medical supplies and hospital personnel. Rosser guessed, "That's the reason the damn submarine that the purser told us about didn't send our tub to the bottom!" Walker commented that the markings may have saved them several times.

Rosser's first confrontation with the war in England came suddenly. He found he had committed a faux pas of major proportions by inviting the young lieutenant and other friends to the hotel for a drink. When he asked the girl at the hotel telephone board to send up a bottle of Scotch and a bucket of ice, he was coldly informed that the Scotch was not available. As an afterthought, the lady remarked that ice was not a luxury of the hotel. Walker's World War I service was to stand him in good stead. He explained that the British did not use ice in their drinks.

A personal trip downstairs by Rosser to see the telephone girl and a pleading explanation of his desperate plight surprisingly softened the attitude of the lady. He was told to return to his room and she would see what could be done. Shortly, an ancient bellboy appeared with the Scotch. Planting a good tip in his hand, Rosser proceeded with his entertaining. Soon he was calling again, requesting that another bottle of Scotch be sent up. At that point, the lady on the telephone froze. Rosser was firmly advised there was no more Scotch. The bottle the old gentleman had delivered to him was the lady's own family liquor ration for the week. Rosser, a little ashamed and embarrassed, decided he was lucky to have had the one bottle.

Sunday morning, the 21st, Rosser was up early. Walker counseled him that he was to wait in his room for the continental breakfast that would soon be served to him, in bed if he desired. Rosser snorted, "Hell, I can't sit around here waiting for breakfast, and besides, I never could eat in bed anyway. I'm going for a walk and look this place over in daylight." To his amazement and consternation, Rosser soon discovered that the heart of the Liverpool business area and much of the surrounding industrial plants had been leveled by the bombing raids of the Luftwaffe. Rosser now remembers that it looked like about forty acres of dust and rubble piled high in the center of the city.

Soon he was back in his hotel room with the announcement that "these sons-of-bitches drive on the wrong side of the road!" He had crossed the street at the hotel corner, he related, and was watching for traffic, "but some bastard driving like a madman coming out of hell from the wrong direction damn near ran me down!"

Walker, being a devout Anglican, urged Rosser to accompany him to church to celebrate their safe arrival. Rosser, feeling that the Lord had snatched him back from tragedy several times in the last two weeks, agreed that "church going may be a good deal." With help from the hotel porter, they located a church within walking distance and attended the Sunday morning services. The stained-glass windows, they noted, had been removed and the openings boarded up. The windows, like other stained-glass treasures of British fourteenth-century churches and cathedrals, they were told, had been carted away to the coal mines in south Wales as insurance against enemy bombs that had been falling on Britain during the past three years.

The Americans were to consider many times how fortunate they had been in overcoming the perplexities of clearing customs, of traveling, and of finding hotel accommodations through the assistance proffered by the Anglo-Iranian Company's representative in New York, B. R. Jackson. Instructions which he had prepared in his letter of February 8 to Rosser were complete:[4]

For purposes of identification, this is to confirm that you are traveling to England under arrangements made by the British Petroleum Representative in Washington in connection with special work to be undertaken by Anglo-Iranian Oil Company for the British Government . . . .

At the port of arrival in Great Britain, we are informed that landing arrangements are under the control of the British Military Embarkation and Landing Officer and that there is a U. S. A. Liaison Officer acting in conjunction with him to whom you should report and who will instruct you regarding transportation to London.

You should also look out for a *uniformed official of Thomas Cook and Son*, who are the Travel Agents handling passenger traffic for Anglo-Iranian Oil Company . . . .

On arrival at the London, or any other main, railway terminus, you should ask for the Thomas Cook representative who will make all necessary arrangements . . . .

Your arrival in London should be reported by telephone to Staff Department, Anglo-Iranian Oil Company, Ltd., Sudbury-on-Thames, Middlesex (telephone number: Sudbury-on-Thames 2900), and you should communicate with that address in the event of any difficulty.

Again, with help of the Cook travel agent, Rosser and Walker boarded the Liverpool-London express at 7:30 A.M.,

4 "B. R. Jackson to Eugene P. Rosser," February 8, 1943, *E* file.

Sir Philip Southwell, ten years later, with his wife and son Richard
in front of Buckingham Palace following Sir Richard's decoration
by the Queen, 1954.

Lloyd Noble.

Frank M. Porter. *Courtesy of Porter family.*

Kelham monastery where American oil-field workers were billeted in 1943.

The market place in Mansfield.

American General Hospital at Mansfield.

Eugene Preston Rosser in 1943.

Donald Edward Walker in 1943.

Gordon Sams.

Wally Soles.

Monday, February 22, the thirty-first birthday of Eugene Preston Rosser. At about 11:30 A.M. the train slowed to a stop outside of what was once the great Victoria Railway Station, the pride of London. The grand old lady now lay in a crumpled heap as if hiding her face in grief.

Upon arrival in London, Rosser and Walker sought out a Thomas Cook and Son agent and showed him Jackson's letters of introduction on the letterhead of the Anglo-Iranian Oil Company. Carefully, the Cook representative scanned the letters, assisted them in getting a taxi, and directed the driver to take them to the Hotel Russell in Russell Square where he said a room awaited them.

As soon as Walker and Rosser could register at the Hotel, they took a taxi to the home office of the Anglo-Iranian Oil Company in Finsbury Circus. As they traveled the detours around great heaps of dust and rubble, they saw that most business houses wore solid board fronts. Here and there they noticed a window that had not been shattered, but an unbroken window was the exception. They arrived in time at Britannic House where the oil company staff welcomed them.

Southwell had come in from the field to meet them, and the afternoon vanished quickly. They were told the office day generally ended at 3:30 to 4:00 P.M., because transportation facilities had been so badly interrupted that considerable travel time now was required by those living in the country and suburban areas. Those living in town required extra time also—but for another reason. The underground railway tubes did double duty as bomb shelters at night. It was necessary to arrive early in the afternoon to get a place in the long lines queued up at the entrance if a person wanted to be assured of a place of safety in the tube at night.

Everywhere Rosser and Walker were seeing and hearing things that brought home to them the stark realization that they now were actually face to face with the awful realities of war. The courage and fortitude obviously displayed by Londoners, after three weary years of war that must have tried the souls of all Britishers, hit both men with great impact. London was a sobering sight and made them feel, as Rosser put it, that they had sure come to the place where help was needed.

Southwell was host at dinner that evening in the R. A. C. Club. Rosser recorded in his diary for February 22, 1943, that they had a "very nice dinner. It was my birthday party." He did not record what they had to eat, but a typical 1943 wartime menu from Simpsons, a London restaurant known to Americans before and after the war, listed such items as boiled salt beef, braised pigeon, jugged hare with port wine sauce, roast duck and apple sauce. Special dishes were grilled pork sausage with braised red cabbage and cucumber sauce, halibut cutlets with hollandaise sauce, and sweets described as barley rolls and marmalade with tea or coffee.

The conversation at dinner centered mostly around the tragedy on the Tejo River where the lost Yankee Clipper had crashed near Lisbon, Portugal. One of the passengers killed in the crash was Harry G. Seidel, a well-known, well-liked, executive of Standard of New Jersey Oil Company.[5] He was a temporary resident of London at the time of his death. Seidel had, ever since the onset of war, been making numerous flights across the Atlantic to help keep oil supplies moving to the fighting fronts. Despite the bombing raids that had damaged his London residence and office, he was returning to London when killed. This tragedy saddened Mr. Southwell who had known him well. Although Seidel was not known by Rosser or Walker, his death overshadowed their knowledge that they had almost been passengers on the ill-fated clipper.

As Walker had noted the early closing of the offices, he also observed the following day the very early hours at which the

5 "Lost in Clipper Crash," *The Lamp*, Vol. XXIII, No. 5 (February, 1943), 4.

office day was commenced. Few taxis or other vehicles were on the streets. They had been fortunate to get a taxi. It was driven by a lady chauffeur who seemed at home behind the wheel. As they detoured through the one-way passageways of the streets that had been cleared of rubble sufficiently to permit the single line of traffic to move, Walker observed that life seemed to be going on as usual. The lady driver's response was, "Yes, we must go on living. Some day Hitler will pay for the 20,000 people he has murdered here in London, and the thousands that lay maimed, crippled, and homeless in our hospitals." At one point the driver casually said, pointing a finger, "Over there is the London tower where Rudolf Hess is imprisoned. After his flight from Germany to Great Britain he was turned over to the authorities by a farmer in the south of Scotland who captured him with aid of a pitchfork."

On Tuesday Rosser and Walker met in the company's conference room with the board chairman and other important directors. This was, at last, the day when the two Americans were given the carefully guarded secret as to where their drilling operations would be located. An area outlined to them on the large wall map in the conference room centered in and around the legendary Sherwood Forest. It lay mainly between the cities of Nottingham and Lincoln. Newark-on-Trent and the town of Southwell, pronounced by most Britishers as Suttle, having, however, no connection with C. A. P. Southwell, were towns near the east end of the presently producing fields. Most of this area they noted lay within Nottinghamshire, and immediately it became clear to the Americans that the producing fields were actually almost in the center of Great Britain.

Rosser and Walker remained in London over Wednesday, February 24, for further meetings with the Anglo-Iranian personnel. Rosser arranged to meet a man from the National Supply Company to discuss availability of supplies and equipment that would be needed from time to time during future operations. Walker spent the rest of the day learning the D'Arcy procedures concerning the keeping of records and reports required by the London office.

By the end of their London stay Rosser and Walker felt ready to pick up the British project. "With the help of our buddies we'll lick the hell out of the Nazis," Rosser commented.

There was real pride in Southwell's face when he added, "We will do it with oil from Britain's own oil fields."

VIII.

The Wheels Start Turning

Rosser and Walker sat together in a compartment of the London–Grantham late afternoon express. The coaches were well filled with the usual commuter crowd. Facing Rosser and Walker were two men and a middle-aged lady intently reading the late afternoon papers.

The dimmed lights spread a soft glow over the seats. The sudden darkness of the winter night seemed to emphasize warnings they had heard in London that they should be suspicious of all strangers. It had been said by someone at their briefings on the day before that even the people living in the immediate vicinity of the producing area knew little of the work going on except it was an important war activity and its whereabouts should be closely guarded. The Americans were told that secrecy was required not only for the important war project but for the safety of the residents of the neighborhood as well.

Three years of war had made the British a taciturn people. The two Americans were being quickly conditioned to the war psychology that seemed to prevail all around them. In guarded tones they speculated on what they would find in the field and in the work that lay ahead in the coming months. The muffled conversation between them abruptly ended when they were interrupted by the train attendant who entered the compartment to draw the blackout curtains.

Mr. Southwell was at the station to meet them. Rosser and Walker were to have dinner and spend the night with the Southwell's at Seven-Mile Post, the country house they had taken as a retreat from the war hazards of London.

Rosser was greatly impressed at the way in which Southwell zipped his small car through almost total darkness on the narrow curving roads leading to his home. Rosser noted that the headlights of the few cars they met on the road, like the car driven by Southwell, were blacked out with metal coverings over the headlights. Only a small beam filtered through a series of pin holes. No lights were seen in the villages and farmhouses of the countryside. The absence of lights and the speed of the small car made one feel that he was being hurled through empty space at terrific speed.

On the following morning after the best night's sleep Rosser and Walker had experienced since leaving New York, they were anxious to get a close-up look at the area. Thus it was that the two Americans got their first view of the D'Arcy Exploration Company offices at Burgage Manor located in the town of Southwell on the early winter morning of February 25, 1943. Burgage Manor, Southwell explained, held its place in history as the home of Lord Byron's mother for several years. She had taken the old house during the years Byron was a student at Cambridge University. Gossip had it that during the summer vacation months, Southwell said, the Manor house was the scene of ribald parties attended by Byron's young college friends and certain ladies of easy virtue who had been invited up from London's Soho district.[1]

The big question uppermost now in the minds of Rosser and Walker was the selection of a place to house and feed the

[1] This anecdote appears in *Southwell, Official Guide of the Southwell Rural District Council*, 8.

boys who would soon be arriving from the United States. Mr. Southwell was preoccupied with the pressing problem of the oil industry on both sides of the Atlantic. He called attention to the fact that the petroleum crisis in Great Britain was such that development of the oil fields of the area had now been given the highest possible priority.

Despite the desire of the two Americans to inspect the several places Southwell had mentioned as possibilities for billeting the roughnecks, Southwell took time to drive them to the field office at the edge of Eakring village to meet the staff. He also took them on a quick trip through the producing area of Duke's Wood, a part of historic Sherwood Forest.

As Southwell showed them the places he considered possibilities, each was carefully considered. The requisition of any quarters Rosser and Walker found suitable for housing the boys would be a mere matter of form. Both Rosser and Walker were opposed to dividing the group for billeting in separate housing. They felt that if the boys could be kept together as a unit, it would be better for all concerned. Such arrangement would help alleviate the homesickness that both knew inevitably would attack some of the boys. The average age of the group was less than twenty-four years. It would also prevent jealousy from developing between the groups and, most important, it would make for an easier job of policing the boys' off-hour activities.

At last they arrived at the Anglican monastery in Kelham village. The monastery was being used as the mother house of the Society of the Sacred Mission and as a theological seminary for the education of candidates for the ministry in the Anglican faith.

After inspection of the premises, the Americans agreed with Southwell that this was their choice as a home for the oil-field workers. There were many reasons that made the monastery preferable to the other available sites they had inspected. Two large bathrooms in each of which were four hot showers that had been installed for a previous military group billeted in the magnificent old building was perhaps the most persuasive reason for their selection. A good rough-neck, Rosser observed, "does not mind getting dirty but damn sure wants a place to wash up."

The monastery contained ample space for the men without crowding. The location afforded comparative isolation from any city or town of substantial size, yet it had the advantages of Kelham village which consisted of a mere dozen or so houses, a post office, and a general store. The area included a four-teenth-century parish church. Across the road only a hundred yards or so from the gateway arch of the monastery grounds stood a pub of ancient vintage by the name of Fox Inn. The entire village and the monastery area were generously sprinkled with towering oaks.

Walker, a member of the Anglican church, presented his church credentials to Father Gregory in the form of identifica-tion from the Bishop of Oklahoma, attesting to his good stand-ing as a member of Trinity Church in Tulsa, Oklahoma. He could not, however, keep from wondering how the robed and sandled Anglican monks of the priory would receive the rough-and-ready breed of the young oil-field workers about to be moved in with them. Rosser's reaction was that with sep-arate quarters partitioned off for the men in one wing of the building, the "padres" would never know they were there. Those working nights, he argued, would sleep most of the day and the day crews would sleep most of the night. "Besides, the work we've got cut out here for us, I'm figuring that not many of them is going to feel like a lot of hell-raising and whoring around in their spare time!"

For Rosser and Walker, several days of hurried activity fol-lowed the selection of Kelham Hall, the name generally used to designate the ancient and majestic building. They moved into Victoria Station Hotel at Nottingham. A 1940 Plymouth car, altered to accommodate eight passengers, was made available for their use. The D'Arcy Company provided ration cards. These did not, however, automatically insure petrol for their cars at all times, since the few remaining distributors in the community often displayed the empty sign.[2]

Walker was immediately occupied with the job of making the quarters ready for the men who would be arriving any day. He contacted various local officials to get a few needed furnishings such as additional bedding, and he lined up a supply of food that could not be delivered until the workers arrived and their ration cards presented to the grocers.

Rosser went to the field to make notes on the work to be done preparatory to the start of drilling operations as soon as the men and equipment arrived. With the help of a few D'Arcy men he started leveling the first well location to be drilled with the new equipment. "Cold as hell today," was the terse entry in his diary of March 1.

Rosser records that Walker talked him into going to church again on Sunday, March 7. They attended Holy Trinity in Nottingham in the morning and in the afternoon rode the bus for tea and dinner with the Southwells. "Best of all," noted Rosser "was a long walk in the woods."

The next day Walker was busy with the local civil defense wardens and the Sheriff of Nottinghamshire. He was arranging for officials to come to Kelham Hall to get the information from each man needed for issuing individual identity cards. These cards, signed by the Sheriff of Nottinghamshire, must be carried on the person of the holder at all times. Walker could almost feel the living presence of Robin Hood.

Rosser and Walker, of course, did not know that at this time

---

[2] Britishers who could show need for the operation of privately owned automobiles were rationed sufficient petrol to drive no more than two hundred miles per month.

H.M.S. *Queen Elizabeth*, converted to a troop carrier, was moving out of New York harbor with a precious cargo of 12,000 soldiers and civilians engaged in war work. Among those aboard were the forty-two young American oil-field workers for whom all connected with the United Kingdom project anxiously awaited. The great ship was surrounded by a convoy of United States destroyers with deck guns at the ready and American sailors standing by. As one of the men said many years later, it sure felt good to see "them convoys" and it was "awfully pretty, too."

The first location to be drilled by the Americans came to be designated Eakring 98. It was numbered 98 because by this time the British had, as Rosser puts it, "wormed" 97 wells into the 2,500-foot producing zone. At this location Rosser continued with the leveling, getting mud pits completed and the lumber matting ready for the rigs. On Tuesday, March 9—a red letter day—he received word from the shipping and forwarding agent through the D'Arcy office that two trucks and several boxes of heavy equipment had arrived on the docks in Liverpool.

At dawn the next day Rosser and little Montie Montgomery, one of D'Arcy's truck drivers, left Nottingham for Liverpool by train. After stops at many small villages, they arrived about noon and contacted the shipping and forwarding agent's office. The next three days were spent in locating and identifying the equipment and boxes labeled for Rosser.

Getting through customs again was trying and time consuming, but loading the equipment was sheer joy for Rosser. Fortunately, two of the International Harvester trucks had arrived equipped with winch and gin-pole attachments. The trucks were about ready to leave the docks when word came that the ship, *City of Edinburgh*, was arriving with additional equipment. Rosser and his helper remained in Liverpool and were able to get the additional, incoming shipment loaded by late Sunday morning of March 14. Rosser's diary entry of that date relates "running up and down the docks all day looking for my stuff, finally found it but no additional trucks which I had hoped for were in this shipment. So we left for Nottingham."As an afterthought he added, "We drove through Washington and Buckston, the most beautiful country I have seen since landing."

Monday, a real English fog was so bad that Rosser and his helpers were forced to wait until it lifted at about 10:00 A.M. before unloading the trucks. By late afternoon the draw works were assembled and the eighty-seven-foot jackknife drilling mast was installed. Rosser's diary of that date observed, "I hope the rest of the equipment and the 'boys' get here right away. These people are really starving for oil. I'm praying we can help boost their production in time to help them a little." Later on the same day he wrote, "If this S. B. Hitler is not laid to rest soon, he is going to mess us up for the next 100 years."

On Tuesday, March 16, Rosser and Walker checked out of their hotel and moved into Kelham Hall. It was a special day for them in another way too. They heard the roar of German bombers overhead. Some of the local people guessed the bombers were headed for targets in Birmingham or Sheffield.

For Rosser their new home was a bit difficult to get accustomed to. "It is really lonesome at night here in Kelham Hall, just Don and me and six cooks," he wrote on Wednesday, March 17. The six cooks referred to were a chef, one assistant chef, two waiters, and two busboys who doubled as dishwashers. All were former British sailors who had recently been released from the general hospital at Mansfield as rehabilitated shellshocked veterans. One of the former sailors characterized their plight: "We just had our ship shot out from under us."

Saint Patrick's Day is hardly a popular day in England. Nevertheless, Walker noticed a few shamrocks. Here and there was a green tie at Burgage Manor where Rosser and Walker had been allotted an office by the D'Arcy Company. Walker had spent the day getting to know the D'Arcy field procedures and setting up a system for the reports that were to be made to Noble and Fain-Porter companies covering the weekly payrolls, drilling information, and a list of other matters to be covered each week in a general letter. Just before closing the office for the day, the big news came. Cook's Travel Agency in Glasgow informed the D'Arcy office that the boys had arrived. Complying with customs, wartime landing regulations, and travel arrangements would delay them 24 hours. The present plan, Cook's office said, was for them to arrive at Newark-on-Trent aboard the Royal Scot at five o'clock the next afternoon.

Rosser finished with all that could be done at the Eakring 98 location until more equipment arrived. He went to D'Arcy's field office and machine shop adjoining the storage yard in Eakring village to confer with the manager concerning work to be done at additional drilling locations and some necessary road building.

Wally Sole, superintendent of field communications for D'Arcy, after all these years says he shall never forget the comment of a fellow employee when Rosser entered the office. Rosser's West-Texas five-gallon hat, leather field jacket, and high-heeled boots were typical of the American cowboy image created by the English cinema. The employee's surprised remark, "Where do you suppose he tied his horse?" has remained fresh in Sole's memory.[3]

Later at Kelham Hall Walker relayed the good news to Rosser that the Americans would be arriving the next afternoon. The day had been cold and wet, but the news brightened Rosser's outlook.

Much to Rosser's irritation, he awoke the morning of the

[3] This incident was recalled by Wally Soles in an interview by the author in England in 1969.

big day with a sore throat and fever. His diary of the 18th carries the entry: "Stayed in bed all morning. Not feeling too pert. Want to be in good shape to meet the boys at five o'clock. Walker says the whole company is going to be on hand to meet them. Boy, will I be glad. The monastery is ready for them. I hope they like it."

Rosser, Walker, and a large delegation from D'Arcy's offices were on hand for the boys' arrival. Montie Montgomery, who was among the company spectators, relates: "It looked like a rainbow to see the colored shirts coming off the train." George Newton, another spectator, was impressed by the cowboy hats and boots.[4]

When Walker saw E. E. Edens climb down with a banjo hanging from a shoulder strap, and another boy with a fiddle case, his reaction was summed up by, "Oh my God!"

All the assurances about the nice, quiet boys given the fathers and brothers at Kelham Hall plagued Walker's mind. He did not know that several carried French harps in their pockets. Walker could have eased his feelings, however, had he known that the banjo, fiddle playing, and music were to become an important interlude in the few hours of daily recreation the boys would have at the Fox Inn. The country music played and sung by the boys for the Britishers, and the English ballads taught them by the Britishers would in time prove to be a real area of international good feeling and cooperation.

With the help of the trucks that had preceded them, the boys and their baggage were delivered to Kelham Hall. All the Americans seemed well pleased with their quarters. The cooks, under Walker's direction, did a fair job of dinner that night with the rations available. From the boys' conversation, Rosser and Walker got the idea that they had experienced a

---

[4] Related by (Lord) George Montgomery and George Newton to the authors when they were in England in 1969.

bad trip. Lewis Dugger described the arrangement in his stateroom on the *Queen Elizabeth*. The forty-two men had been assigned ten staterooms, enlarged to accommodate four passengers each. One man showed up in each of two of the staterooms without a bunk. Fortunately, all of them had been given a bed roll on boarding the ship and in typical oil-field fashion, the man who was to sleep on the floor in each of the two overcrowded staterooms was determined by a roll of the dice that some obliging member of the group handily produced. Dugger's memory of this incident, he says, is good because he was the loser in his stateroom gamble.

Friday, as previously arranged by Walker, the Sheriff of Nottinghamshire and his deputies came to Kelham Hall. The day was spent in fingerprinting, filling out the rationing forms, and cataloging personal information required by the authorities for preparation and issuance of individual identification cards. Rosser remarked, "Don is as busy as a cat on a tin roof getting the boys signed up and registered with the police and ration board."

Saturday, March 20, was a day of adjustment. It was market day in Newark-on-Trent about two miles from Kelham Hall. The boys were there to the man. By noontime they were circulating freely through the town's cobblestoned square and putting on a good show for the English farmers and their customers at the stalls where meager farm products were being offered to the public. Rosser remained in bed with a sore throat, but he remembers the boys' excitement as being like a bunch of high school boys at a county fair.

Mr. and Mrs. George Newton, who then occupied one of the market stalls as they do today, vividly remember the arrival of the boys. "It was a great day," they will tell you. The boys mingled with the English farm folk and became customers of the merchants around the town square area. It did not take long for them to discover Mr. Curry and his bicycle shop. The dozen bicycles in stock were snapped up at once. The additional four or five wheels Mr. Curry was able to secure from other mechants were sold as quickly. The boys, pairing themselves, bought the bicycles in partnership. The sixty-five U.S. dollars each had been given on embarking from New York certainly came in handy.

Mrs. Newton's memory of the white Stetson hats; the colorful display of red, yellow, green, and pink shirts, with an occasional plaid; and the embossed high-heeled boots worn with one trouser leg tucked in, created a glamorous picture. "They made me think of my flower garden in spring," she says. The boys laughed off such frequent questions as "Where is your gun?" and "What are you going to do in England?" Some resourceful member of the group started the confidential rumor which caught on quickly that they were in England to do a movie. Mr. Newton also remembers one of the surprising habits of the boys was mounting their bicycles from the left side instead of the right as is the common practice of the English. They returned—about half of them on bicycles— late in the evening to Kelham Hall fully relaxed by their day of rest and fun on market day.

On the whole, it had been a great get-acquainted day. J. W. Nickle and L. B. McGill had bought English walking sticks— for what purpose no one knew. Ray Hileman met George Newton that day. Their close friendship stemming from their common interest in gardening was to be a boon to the group in the coming months. Al Morton found a glass-domed clock in a shop and bought it. En route to Kelham Hall on his bicycle, he dropped it and smashed the glass dome, but that clock was to start a collection that has now grown to more than one hundred quaint clocks. In time he came to own a number of fine specimens of this particular type of clock.

After a few beers, De Havely tried to put through a long-distance telephone call to Stroud, Oklahoma. Finally after many long sessions with the operator, he got Stroud, England. "Hell no, honey, I want Stroud, America," he yelled. "Not Stroud, England, or Stroud, South Africa, or Stroud, Australia —Stroud, America, I'm telling you!" After this, De Havely was and is now known to the other boys as Stroud America.[5]

Sunday morning, March 21, was Don Walker's birthday. Rosser's cold hung on so he remained aching in bed all day. Two others complained of sore throats. Although Walker spent most of the day doctoring the ailing, he and Rosser, nevertheless, had cause to be thankful on this rainy Sunday. The boys had arrived safely; there was no serious illness. All were registered. Identity and ration cards had been issued and all were settled in comfortable quarters. Everybody was anxious to get on with the job.

Monday was still wet and cold. The boys were disappointed that more drilling equipment had not arrived. Rosser knew that healthy boys could not be kept idle. They must be put on some sort of work. The areas already drilled by D'Arcy needed a good cleanup. The storage yard could use a couple of good pipe racks. Rosser took over one of the company's A. C. rigs (alternating current motors). J. W. Nickle as driller and his crew of derrickman Gerry Griffin, helper Little Joe Webster, and motorman Glenny Gates, were assigned to the morning tour of twelve hours. Horace Hobbs, as driller, and his crew of derrickman Ed Boucher, helper Al Morton, and motorman John McIlwain, were given the afternoon tour of twelve hours. Both Rosser and Walker felt that Project U K was off to a start.

In the meantime, back in the United States, Ed Holt had received a copy of a cablegram dated March 17 to Anglo-Iranian's New York office that the boys had arrived in England. Holt immediately relayed the news to all the families of the boys. It was a great relief to Noble, Porter, and Holt to know the group for whom they felt responsible was safely landed.

[5] Reminiscences of Gene Rosser and Don Walker, 1968.

Also in the meantime, Walker had found time to assure Holt that the quarters were adequate and comfortable. There were eight shower baths, two men to a room, except Rosser, Walker, and Sams who each occupied a single. All rooms had corner fireplaces. The lounge and recreation room, dining room, and kitchen had central heat from a fuel oil furnace when oil could be secured.

As time went on, it was felt that Kelham Hall was right in every way for the boys. It was apparent that the monks and the boys were compatible. Their relationship evoked puns from outsiders such as, "Let's go to Kelham Hall and see what mischief the Rogues and Robes have been up to."

The rigid rules of the monastery appeared not to make the boys rebellious. Instead, monastery rules were respectfully observed by even those boys who, when once outside the monastery, sometimes drank too much and had to be quelled by the police in Newark, Southwell, or Nottingham.

As for the monks, before long they became fond of these young, brash Americans. Their spokesman, Father Gregory, the prior of the monastery, eulogized the Americans in an article he wrote for *Naft*, the Anglo-Iranian Company's publication:[6]

Kelham is a pleasant but undistinguished village in Nottinghamshire. It contains a few houses and cottages, a general store and Post Office combined, a public house, a fourteenth-century church, and our place.

In due time, the property was acquired by the "Order" and our community moved in. We added a first wing that runs at right angles to the house; then a second wing at right angles to the first; and lastly, a chapel to make up the fourth side of the courtyard.

The house, besides being the Mother-house of the Society of the Sacred Mission, is a Theological College. When the war started there were about 150 students here, as well as some 25 members

---

[6] Father Gregory gave the author an undated reprint of the article when he visited with him at Kelham in 1969.

of the Society, but, since then, our numbers have been gradually dropping, for, although the students in residence at that time were reserved, newcomers were not, and they went off in the Forces when they reached the age of eighteen.

As we shrank in numbers, we retreated into the main building and the wings were boarded off to house, first, soldiers and then sailors; but about a year ago the latter also went away and the wings stood empty—but not for long.

One day we received a visit from Mr. Southwell, who wished to use the wings as a hostel for some Americans. We were a little alarmed at this. True there were to be only some fifty of them as against the two hundred soldiers and sailors who had been there during the previous three years, but when the question arose as to the noise and disturbance of our monastic seclusion, the odds seemed in favor of the newcomers.

We expressed our alarm, and received a sympathetic hearing, but were assured that our fears were groundless. Now it stood to reason, we were told, that these men would be as determined as we were to keep the place quiet. They would not want their sleep disturbed and they would take care that none of their own fellows offended in this. This seemed reasonable, and our alarm was dispelled.

Then the decorators took over. Two rooms were converted into kitchen and larder; others were fitted up as dining-room, lounge, and recreation room; and the rest of the rooms were furnished as bed-sitting rooms, each to take two or three men; this last is what they were built for and what we had used them as.

The first arrivals were the domestic staff, consisting for the most part of sailors invalided out of the Service. Then a few days later the Americans arrived. (Walker and Rosser) I have probably seen as much of them as any of us have, for I was what they call "the Contact-man." I met them on arrival and was in their part of the building a good deal of the time during their first few days. These were passed mostly in being lectured; I attended in the hope that I should hear a lecture on "British Corns and How Not to Tread on Them," but they must have had that one earlier; all I got was Air Raid Precautions. Though our guests have been with us some months now, they have not yet had to put their knowledge of the latter into practice.

There was a certain amount of shyness on both sides at first, but that soon wore off and now we mix and talk quite freely. In these inter-Allied conversations I have one great advantage. The Americans all knew a good deal about this country before they came here. Apart from the general knowledge that they have of us, similar to that which we have picked up regarding them from history lessons, newspapers, and novels, they had seen a lot of our deliberate publicity—and not merely *Mr. Chipps* and *Mrs. Miniver* and such-like fantasy; there are many publications put out in the States that give a real picture of England at war, and some of these were lent to us by our guests. Again, they have all been lectured on our insular pecularities; they won't admit that they have been warned of certain topics, but just try them on the British Empire and there is a polite silence. But whatever they have read or heard, they still find us a bit odd; not that they say so, but I have a feeling that they would agree with an Australian to whom I was talking recently—his term of greatest condemnation was "quaint" and he seemed to apply it to most things British.

Now, my great advantage over the Americans in the matter of mutual understanding is this: I have been to the cinema. I am not going to insert here a digression on "Visual Aids in Education" or some such, but it really is most interesting, and at times almost uncanny, to see how like the films they are. For one thing, like the people on the films, they dress well and seem to justify the boast that you cannot tell a man's social status from his clothes. Bright-coloured suits, broad-brimmed hats, or caps with peaks pointing to heaven, and high-heeled boots seem to be the favourite dress when not working. The local children, and probably not the children only, were genuinely surprised to find that the Americans did not ride horses and let off revolvers about the place. But where I scored most heavily was in the matter of speech, for they talk the language of the films perfectly. I had no need to say "Sir" to them as they had to me, I felt that I knew them well at once; there is nothing odd about them, I had seen and heard them often before. Add a nodding acquaintance with American magazines—especially the advertisements—and you feel that you have known them all your life.

The assurances about quiet were fully realized in practice. I hope we have disturbed them as little as they have disturbed us.

Our guests have put Kelham on the map, judging by the number of their visitors. As is natural, American visitors are interested, so they come to see us; they call for a meal or perhaps to spend the night in the hostel, and they are introduced to us. Most seem to find something piquant in this close proximity of such typical Americans with a British religious community. Witness, for example, the film that the Gainsborough Company made of these tough guys against a background of our people. Some of these callers have heard of us before, or have seen pictures of our chapel and want to see the actual thing.

To others it comes as a complete surprise; but all wish to look around and are duly impressed. I have shown numbers round, but the most gratifying occasion was when I was not doing the showing myself. There was an American General who had come out of his way to see the place that he had heard so much about. He brought his entire Staff. They "did" the chapel at top speed, but the "Brother" who was showing them round was not to be done out of his full story; the faster they walked, the faster he talked, till one of the officers said to me in a hushed voice, "Gee, some salesman."

"I like them—the Americans" Father Gregory declared, "I am glad they have been here. I hope that they take back to the States memories not only of our weather, but also of a village in Nottinghamshire, a little shaken by their coming, but soon relapsing into its normal undistinguished pleasantness."

IX.

"I Say, Old Man"

In the English Midlands, the late March rains deluged the countryside. To make matters worse, the cold winds off the North Sea stung the face and drove the cold to the bone. Roads in the field and around the drill locations were knee-deep quagmires that made movement into and out of the field slow going. The boys, however, had long since been initiated into the hardships that go with the life of an oil-field worker. They were, indeed, a rough-and-ready breed of men. They had grown up taking the weather. Rain, snow, blazing heat, or the sandstorms of the American Southwest were part of the job. The roughneck took it as it came and as a matter of course.

The D'Arcy rigs were equipped for wartime operations. Telephones with loud-sounding electric gongs were considered a necessity.[1] This type of communication gave the D'Arcy Company immediate access to the drilling crews in case of air-raid warnings and likewise the all-clear signal. Rig telephones were used also for drilling reports and other related field information.

Work on the rigs was carried on under many handicaps. The

[1] Clause (d) of the official contract entered into by Anglo-Iranian with the Noble and Fain-Porter companies provided for the telephone equipment and gongs. See O file, George Otey, Sr., Ardmore, Oklahoma. This was verified in 1968 by the author in an interview with C. W. Soles, head of the Department of Field Communications for D'Arcy in 1943.

lighting arrangements were perhaps the most difficult. Two small shaded lights at opposite corners of the derrick floor were permitted. One similar light served the doghouse and another light was located near the mud pumps; such was the rig lighting system. Lights were heavily shaded so as to direct the light beams in a small circle on the rig floor. Wartime regulations limited the lights to one candle power for each foot above the floor on which it was placed.[2]

Following the first twelve-hour morning tour on the D'Arcy A. C. rig, J. W. Nickle, the morning tour driller, routinely reported the 1,010 feet of hole drilled on his tour. The D'Arcy man at the field office suggested Nickle should take another measurement. The footage drilled, as reported, could not be correct. No such drilling record had ever been reported by a D'Arcy drilling crew. Nickle assured him it was accurate. A few minutes later, a supervisor in the production department at the D'Arcy field office rang up to say that something must be wrong. Surely Mr. Nickle's team, as the D'Arcy men referred to drilling crews, had not been able to drill 1,010 feet in one tour. Nickle, however, was emphatic. The measurement was correct, he declared, and went back to his business of feeding the drill pipe into the hole as the rotary bit cut its way into the comparatively soft formation. Serenity of the operation, however, was not to last long. Again, the loud sound of the telephone gong rose above the rig noise. There was an inflexible order to all drilling crews that the telephone gong must be answered immediately, notwithstanding any other operations at hand. Nickle, assuming that another D'Arcy man was going to take issue with his report, with rising irritation switched off the power and applied the rotary brake before answering the insistent ringing of the telephone gong. The man on the line

2 Information on lighting arrangements was furnished the author by Sir Philip Southwell and verified by Gordon Sams, Tulsa, in December, 1970. Sams was the tool pusher for the four American rigs used in the English project.

was Sandy Bremner, works manager. Nickle listened as the latest comment concerning his report came through the receiver.

"I say, old man, your drilling report just can't be true." The speaker got no further. By this time Nickle and his crew, who were now standing by, really had their blood pressure up. In typical and forceful American oil-field vernacular, Nickle told the big boss he was damn sure of his figures. They had been carefully checked. Bremner wanted to know how many bits had been used and called attention to the English practice of changing bits at thirty-foot intervals.

It was Nickle's time to explode. "What the hell has changing bits got to do with it?" Why should a bit be changed as long as it was making holes? The bit now in the hole had cut more than sixty feet of rock. As a parting shot before slamming up the telephone receiver on its hook, Nickle yelled to Bremner, whom he had never met, "If you're so damn sure that we are wrong, why in hell don't you smart 'sons-a-bitches' get out here and do your own measuring?"

Having been so brashly and unexpectedly rebuffed, Bremner appealed to Rosser, "If this fellow Nickle's report is correct, the crews should be slowed down before they wreck the equipment!" Rosser assured Bremner he had full confidence in Nickle's crew. The figures reported were in fact not surprising to Rosser. He had on many occasions seen more footage drilled in the same time and was accustomed, as the crews were, to not wasting time and bits by making changes according to any footage or time schedule. As long as a bit was satisfactorily cutting hole, it was Rosser's policy to keep the drill stem turning to the right, a common United States oil-field expression to indicate the driller was drilling ahead. The footage, in fact, was nothing more than estimated by Ed Holt months before when Southwell was wanting to purchase ten rigs in the United States.

Rosser placated the company men with the assurance that Nickle and Hobbs knew what they were doing and suggested that they let them go ahead on their own. Why waste time changing bits as long as they were doing the job? Rosser explained the drilling crews' main objective was speed, and that the changing of bits depended on many factors: weight of drilling mud, the degree the hole might be off vertical, and, of course, the formation being drilled. Nonchalantly, he told Bremner he would take full responsibility for the work of the crews.[3]

Finally, a count of the drill-stem joints convinced the D'Arcy production men that Nickle and his crew had actually drilled the footage as reported. Within a few hours the gossip of the incredible speed being made by the American drilling teams had gone through both the field office at Eakring and the Burgage Manor office. Unbelievable as it was to the men of the production department, including Southwell and other D'Arcy Company officials and the people of the great Anglo-Iranian Company at Britannic House in London, the daily drilling figures of the American crews soon became pleasingly acceptable to them.

Rain and more rain—rain every day for a week did not help the roads in the field areas and around well locations. Laying fuel and water lines was a nasty job. Hauling scrap into storage yards in a downpour certainly was not an exciting operation. But the boys must be kept busy. Most of them had found the comfort of the Fox Inn and the relaxation they got from a few beers. "No danger of overdoing the beer routine," one of the boys wrote to Ed Holt, "because beer is rationed to the Pubs here." Most of the time the Fox was out of light beer and ale.

---

[3] The clash between Sandy Bremner, works manager for D'Arcy and J. W. Nickle, American driller, whose crew drilled 1,010 feet in one tour (twelve hours) was told to the author in detail by Rosser in 1970. Nickle's skill as a driller, coupled with his drilling experience, is attested to in his personal history record, *E* file.

"This black bitter stuff is sure named right. They call it 'Bitters.' The Pub usually has plenty of it, but who wants the stuff—unless you was raised on it. A little salt helps kill the bitter taste," he wrote.[4] Binty Hayward, daughter of Captain Hayward, owner of the Fox Inn, remembers how surprised her father and other natives of the village were to see the boys pouring salt into their beer and ale.[5]

Rosser reported on March 26, "Everything running smooth." Everyone was now working except a couple of boys who still had sore throats, and Pete Oaks, who had the first accident on Friday, April 26. Oaks had dropped a joint of three-inch pipe on his foot, breaking three toes.

Next day, the boys got a gin pole rigged up on the D'Arcy caterpillar tractor. They were also able to use one of the American trucks equipped with a winch and gin pole for moving a Wilson pump from the Eakring yard to Farley Woods number-two location. The D'Arcy men were astounded at the ease with which two men loaded and unloaded the heavy pump. The self-loading truck was the first any of them had ever seen and was perhaps the first such truck in England. Self-loading trucks were not to be found, even on the deep-water docks. Only about ten minutes were required to load the Wilson pump. This speed factor and the saving of so much man power effort was almost as amazing as the drilling speed the Americans were getting out of the big English A. C. rig. The boys had taken the big A. C. rig with 134-foot derrick designed for drilling depths of eight to ten thousand feet, on Monday, March 22; and on Sunday, April 4, twelve days later, Rosser noted, "We are drilling in today with the D'Arcy rig."

In the meantime, Rosser and the truck drivers had been busy. News was received Saturday, March 27, that two Inter-

---

4 "Ray Hileman to Ed Holt," April 6, 1943, *E* file.
5 Interview with Mrs. George Lloyd, nee Miss Binty Hayward, Loughboro, England, in the summer of 1969.

national Harvester trucks and 171 boxes of equipment had been received on the docks at Cardiff. On Sunday afternoon, Rosser, Robbie Robinson, and Luther McGill had left Nottingham by train to get the trucks and other material.

The trips to Cardiff and other ports afforded Rosser and his fellow workers ample opportunity to view the increased war supplies and equipment piling up on the docks along the west coast. The reason for the pile-up, they surmised, was related to the activity in North Africa. One heard snatches of war talk at the docks that proved more accurate than some of the radio news accounts of the war's progress the men had access to in the monastery.

Following the November landings of Allied troops in North Africa, a new build-up of supplies and material preparatory to the invasion of Sicily and the mainland of Italy had begun to pour into the ports on the English south and west coasts.

Even at this early date, some of the equipment, they noticed, was marked for retention in England. All the south and west coast ports were being taxed to capacity by the daily arrivals of armament and military supplies. The army and navy food build-up being received on the docks was especially noticeable. Oil and aviation gasoline storage was being filled and additional storage being built at several south coast ports.

On March 8 Rommel had driven back General Lloyd Fredendall, commander of General George S. Patton's 11th Corps, at the Kasserine Pass. Rommel, by striking at Montgomery's 8th Army in Tunisia the first part of May, attempted to follow up his advantage, but after a second panzer clash with Montgomery's tanks with devastating losses to the Germans, Rommel went home because of ill health and never again returned to his African Corps.

For good reasons security regulations at Cardiff were tight and rigidly enforced. Most of Monday, March 29, was spent in convincing port authorities that Rosser, as a consignee with his truck drivers, was entitled to go on the docks to examine shipments before accepting delivery. With the help of T. T. Passcoe, the freight forwarder, the Americans were given passes to the unloading area. Rosser's diary of March 29 notes on that date, "We finally got a look at the trucks, one International K-7, and the big International K-8—sure looked good to me." But more time-consuming war regulations were to come. Before the trucks could be loaded and put on the road, the requirements were frustrating and irritating. Two British licensed truck drivers with trade plates (license tags) must be used to move the trucks off the docks. The trucks could re-enter the docking area for loading of the material as war industry trucks. By 9:00 A.M. of the next day, two truck drivers arrived from London with the required trade plates. Rosser records that he and the boys "had one hell of a time getting a can to bring petrol to the pier." Finally, he had to get a representative from the Cardiff office of the Ministry of Petroleum to obtain containers and bring petrol to the docks. Rosser, fretting at what he deemed a waste of time, joined the customs officials in opening up and checking the crates and boxes of materials to determine if everything checked with the ship's manifesto.

To Rosser's surprise and elation, they found that one of the boxes of spare pump parts contained a box of most welcome cigars. "Really stogies," says Rosser. The customs official pounced on the item with great glee and much grave hand rubbing. He intimated that Rosser was in big trouble and might be held as a smuggler.

At last all the items were cleared and loaded. The trucks' self-loading equipment had cut the time to about one third the time that would have been required by use of the overhead traveling cranes at the docks. Petrol coupons had been secured for the return trip to Kelham.

Only the matter of the contraband cigars remained to be settled. The innocent-looking cigar box had been moved to the customs office where it was being held in the custody of the chief customs officer on duty. Rosser and the boys were anxious to get started. It was now late afternoon and the best they could do was to get the trucks off the dock and through the well-guarded gates of the high steel fence surrounding the dock area in order that they might be ready for an early start the next morning. The trucks had not yet been equipped with regulation headlight dimmers and must be moved during daylight hours or be subject to the hazards of traveling the narrow, curving roads without lights. The trucks might also be impounded and taken off the road by some defense warden.

With his usual aggressiveness, Rosser rushed into the customs office to get his cigars and inquired as to the procedure. Gravely, the man in charge announced he must pay the existing tariff and the special war tax applicable to tobacco in any form brought into Great Britain for resale. He explained that the cigars were so classified because the box had not been opened, and no part of its contents used prior to being searched by the customs officials. "Well," Rosser wanted to know, "How much do I owe you?" The customs official examined his book of tariff regulations at length and finally explained His Majesty's tariff, tax, and the penalty levy amounted to so many pounds, shillings, and pences. "But how much U.S. money?" Rosser wanted to know.

"Well, in your money," the official advised after consulting printed tables in his book, "that will be $32.25." The amount named was several times the retail price of the best cigars in the United States.

Rosser needed only a fraction of a second to give the customs man an answer, "Listen, buddy, you call His Majesty, the King, and tell him what he can do with these cigars! I hope they won't be too rough for him, and I'll leave them right here for his use."

Rosser contends to this day that the customs chief, without

the slightest signs of effort, rose straight up into the air some two or three feet. He came down with both arms flung out wide and his eyes ablaze, shouting, "I say, old man, you cahn't say such things about His Majesty. You have insulted the King of England!"

Taking considerable youthful satisfaction out of the customs official's apparent shocked displeasure, Rosser rushed out before more trouble could develop and mounted the empty seat alongside Robbie, on the big K-8 truck, motioning McGill to follow with the smaller truck. They plowed through the open gate without stopping to be checked by the guard and on to the Queens Hotel in downtown Cardiff, where they found food and lodging for the night.

They were up and out of Cardiff long before the 7:30 A.M. breakfast hour. Rosser relates, "Everything seemed to be going our way when all hell broke loose in the big truck power take-off gearbox. The 'Tulsa' winch had kicked the worm gear out. I was driving, pushing the big boy pretty hard, I guess. The noise sounded bad. I was scared to death that the winch assembly was ruined."

The two trucks returned four miles to the Arlington garage at the edge of Cardiff. Rosser and the boys went to work. At 9:00 P.M. that evening, the damage was repaired sufficiently to put the truck on the road. Nothing was broken; the bronze bearing was frozen to the shaft. With the help of one of the Arlington garage mechanics, they were ready for another early start the next morning.

At 6:30, April 1, they left Cardiff. One of the boys remarked that they had already had their April Fool's trouble. About 10:00 A.M., a "Fish & Chips" sign was sighted. Having missed breakfast and had no dinner the night before, they were ready for food. The lady in charge of the small eating place said there were no fish and chips to be had, but she could make them a toasted cheese sandwich and a pot of tea. McGill

115

hastened to assure the lady, "It don't make no difference. We'll eat anything that don't bite us first." The long trip to Eakring ended at 7:30 P.M. With a pocket flashlight, Rosser noted that the big truck's speedometer showed 225 miles. After a hurried dinner of leftovers, Rosser and the boys fell into the inviting beds of Kelham Hall, content with the knowledge that they had brought home the bacon. They now had their trucks and enough equipment to finish rigging up two National 50 rigs.

The big K-8 was capable of handling loads up to fifteen tons. The big boy was destined to become a real part of the U. K. Project blessed with a personality all its own. The D'Arcy men marveled at what it could do. When the driver put the old boy into low gear, absolutely nothing could stop him, they said. Just who christened the big fella is not known, but soon everyone, D'Arcy men and the Americans alike, knew Ole Lo-Go as well as they did any of the boys. They held the truck in great respect for the mighty feats of strength that it exhibited almost daily.

Twenty-five years later, on one of those rare days in June that come to the English Midlands, little Montie Montgomery would reverently stand in the storage yard near the Eakring office of the D'Arcy Company and tenderly caress the rusted chassis of the old truck that had long since lost its heavy coat of camouflage green to the salty winds of the North Sea that whistle through in winter. Montie was heard to mutter to a few of us who had paid a surprise visit to the Eakring office, "If Ole Lo-Go couldn't do it when you gave him the low gear, then no truck could do it." He continued, "We have been intending to clean up the yard and take Ole Lo-Go's chassis to be smashed up for junk metal, but no one seemed to want the job." As we left the yard, a backward glance at the last remains of Ole Lo-Go, leaning there against the steel fence post, made one think of the bleaching bones of some prehistoric monster.

Now that the boys were into the month of April and had on

hand the equipment to rig up two of the four rigs originally destined for the project with two trucks on location, Rosser felt pleased. At least they were making headway. If the balance of the equipment was received within the next thirty days, the completion of the one-hundred-well program within the one-year work period could easily be accomplished.

Robbie and McGill spent the day, April 2, working on Ole Lo-Go's gearbox and winch. Rosser's diary records: "They reported it was now as good as new—thank goodness."

During the next few days, the boys finished assembling and rigging up the two National 50 rigs complete with the jack-knife mast. By Monday, April 5, they were ready to join the British project drilling operations.

One detail remained to be completed. The rigs, including the mast, fuel, and water tanks, the trucks, and all other equipment must be painted a shade of vivid green that matched all the equipment of the D'Arcy Company even to the office, garage, and outbuildings. The shade of green seemed to blend perfectly with the spring pastures and the new foliage that was coming out in the countryside. The solid green camouflage merged so perfectly with the spring foliage, that none of the oil field equipment could be distinguished from a plane at three-thousand-foot altitude.

Rosser felt that he was making progress. He and Don Walker wrote the home office that there was really no advantage in air mail as most of it was transported across the Atlantic by surface carriers. They specified that certain magazines and periodicals be sent the group as well as a daily newspaper, particularly, the Sunday edition of the Tulsa *World* in order that the boys could keep up with "Dagwood" and other features of the comics. They mentioned that they had a miniature billiard table for the recreation room of their quarters and bridge tables. They suggested that baseball, softball, and badminton equipment be sent over for outdoor recreation.

Gene wrote that the whole setup, living quarters, was better than he expected. He said he had not lost any weight but had eaten a lot of potatoes and Brussel sprouts, "had only one egg so far and no steak." He added: "You need not have any fear in sending a drunkard over here. It would be a good place for him. You can't buy enough whiskey to get drunk on and the beer is like soapy water with no alcohol in it." He mentioned that there had been a couple of alerts and gunfire in Nottingham but no damage, although "you can look around some of these towns and see that there has been plenty of excitement at one time." He suggested to Ed Holt that if either he or Lloyd Noble came over to visit the project they should bring two watches—one for Mr. Southwell, the other for his use, as watches were unobtainable in England.[6]

[6] "Donald Walker to Ed Holt," March 6, 1943, and "Eugene Rosser to Ed Holt," March 16, 1943, *E* file.

X.

"What Goes Up When the Rain Comes Down?"

Within a week after the drilling crews arrived at Kelham and before American equipment was assembled, Rosser and Walker advised Ed Holt of activities under way. They reported travel time for the crewmen that ended at 12:01 the morning of March 22 when the men were put on the payroll for twelve-hour shifts, with the exception of truck drivers who would work on an eight-hour basis.

Most of the men were employed in unpacking and cleaning equipment on hand, racking casing, and were engaged in other tasks necessary before all efforts could be directed toward drilling operations. Drilling crews took over a D'Arcy rig at noon Monday, March 22, when it had drilled to a depth of 175 feet and by midnight Saturday, March 27, it had reached 1,565 feet. All were anxious to get on with the job for which they were employed. The men who had recently arrived were recovering from the series of colds, the laryngitis, huskiness, and sore throats attributable to changes in climate and weather conditions. Don wrote that during the period Kelham Hostel could easily have been called Kelham Hospital.[1]

The rains seemed to have slowed in early April, and the early morning fog had given way to the warm spring sunshine. The towering trees, shrubbery, and the hedgerows along the

1 "Rosser, by Walker, to Holt," March 29, 1943, E file.

stone fences were putting on their lush green foliage. One who has witnessed the great variety of color that comes to Lincoln and Nottinghamshire in the autumn may think that it presents one of nature's unexcelled pageants. But the early spring has its own show of magnificence. The yews, chestnuts, copper and black beeches, the tall poplars, the variety of white and red English oaks, the pines, and the spreading lindens and cypress all contribute their particular early spring shades of dark red and lemon greens in a breathtaking prelude to summer. The beauty of these trees is enhanced through a careful program of reforestation by the landowners.

The warming days and a bit of sunshine were a potent tonic to the boys. There was an eagerness in their friendly competition as to which crew drilled the most hole during the week. The glories of the unfolding spring would have been enjoyed much more by Rosser if the mail from home he was looking for would come through. He had now been in England since February 20, and he had received only one letter from home. At last, on April 8, the hoped-for mail came in. A letter from Bernice assured him that everything was fine at home; the baby had been ill, but nothing serious. She was now fully recovered. Rosser's spirits rose with the receipt of the letter. But the boys were to be bothered throughout their stay in England by the delayed mail deliveries. Each letter and package must filter its way through the censor. One bag came in showing plainly by the yellow marks that it had been water-soaked at sea. Business communications with Rosser and Walker from the Noble and Fain-Porter companies were addressed to them in care of B. R. Jackson, New York. The delivery of this type of mail to the Anglo-Iranian offices in London and the D'Arcy Company offices in Southwell was much faster than individual mailings. Rosser's late evening diary entry for April 8 reports he "did a little better at poker tonight, won 10L/10S," which he estimated compensated him for the losses of the night before.

It seemed that with the good, there was always a sprinkling of trouble to season the busy days. A note from Edgar Holt[2] had come saying Jackson was advised that one of the vessels carrying equipment for the British project had fallen prey to enemy submarines and had gone down carrying:

> 1 Drilling Mast
> 1 C-150 Slush Pump
> 1 C-100 Slush Pump
> 1 Traveling Block
> 1 Tubing Block
> 1 4½" Rotary Drill Stem; string of 2,500 feet
> 1 3½" Rotary Drill Stem; string of 2,500 feet
> Substitutes
> Swage Nipples
> 1 Box of spare parts for Slush Pumps

Another letter, dated March 6, had arrived from Noble.[3] It had been in transit just a month, Rosser noted. Although marked personal, the letter came via Jackson's office in New York. It would have been sent over with the boys but arrived in New York too late and had been forwarded by surface mail. The letter, addressed jointly to Gene and Don and afterwards made available to the boys, made all of them feel the importance of their mission and the pride in being a part of the Noble and Fain-Porter organizations. Noble observed:

> I am sure that when the record of this project has finally been written, it will be one that will not only be a credit to the organization, but it will be one to which you can look back to with a feeling of great satisfaction and pride as having had an opportunity to make a real contribution toward shortening this terrible conflict in which we are now engaged.

2 "B. R. Jackson to Ed Holt," March 30, 1943, *E* file.
3 "Lloyd Noble to Gene Rosser and Don Walker," March 6, 1943, *E* file.

It is my firm conviction that no soldier on the front line of battle has a more vital assignment than that which is being carried out by the boys under your supervision and direction. I had an opportunity to see and talk to the boys before they left, and stressed the various things that we mutually discussed before you went over. I pointed out not only the vital importance of carrying on our drilling assignment in a competent and efficient manner, but also we expect everyone to make themselves as nearly as possible a member of the community in which they live.

There were quite a number of happy faces around here the day we heard from New York that you had arrived safely in England. The papers had just carried an announcement a big ship had been lost. As soon as we got the good news of your arrival, we, of course, immediately contacted Mrs. Rosser, and as fast as I have been able to get around to it, I have tried to tell all of Don's girls that he is OK. But, of course, I probably won't get around to all of them, as that is a pretty big assignment.

There is not the slightest doubt in my mind but that they [the boys] will justify every confidence that we have placed in them, and that I know that we and their country will be justly proud of their contribution. I just don't have the command of language to attempt to draw a word picture of how much it will mean to all of us, when we consider the millions involved, if we can feel, after this war is over, which we will certainly win, that as a result of this effort we have had a part in reducing the duration of this awful war one minute, one hour, or one day, and that we have made a real contribution to this end.

Walker took great pains to see that all the boys had an opportunity to read Noble's letter. It was shown also to the brothers of the monastery. By this time the friendly, courteous association of the monks and the boys was beginning to manifest itself in a way that would bring harmony and good will to this incongruous group of men who were to live under the same roof for the next year. Father Edgar had commenced a delightful routine of meeting the boys coming off tour and saying good-by to the boys going on tour. He always had a riddle that made them smile through the work hours, such as:

What goes up when the rain comes down?
        An umbrella!
What is the first thing you put in a garden?
        Your foot!
What is round as a biscuit, busy as a bee, and the prettiest little
thing you ever did see?
        A watch!

Another remembered by the boys: "What did the pig say when a man got him by the tail? The pig said, 'This is the end of me.' "

The monks made sure that a schedule of Sunday services or other special services in the chapel was posted in the boys' wing of the monastery.[4] Although Walker was the only regular attendant at the services, the monks continued posting the notices each week.

Much to Gene and Don's elation, the monks and the oil-field workers got along famously. An attaché of the American Embassy in Washington later remarked that some of the people in the Embassy were making bets as to whether the boys would take on the robes of the monks, or the monks would be absorbed in the oil-field operation.[5]

Only once does Father Gregory remember it was necessary to complain about the boys' activities. The heavy trucks were being driven into the courtyard of the monastery, resulting in badly cutting up the hard surface driveway from the gatehouse into the courtyard. The boys immediately brought in materials and resurfaced the driveway. A Kelham villager remembers one of the boys observing with a chuckle, "The damn road had needed resurfacing for twenty years!"[6]

[4] Don Walker's scrapbook contains some of the church schedules or programs of the era.
[5] In the *E* file, there is a letter dated July 24, 1948, from Smith Turner, the attaché, with the enclosure of an undated article he wrote, where reference is made to this, "My Guide in Boots and Stetson."
[6] *Ibid.*

Gene and Don had thought that with the problem solved of getting the housing for the boys satisfactorily arranged in one group so that the combined food ration stamps might be used as a unit for purchase of staples from the package stores, and fruit and vegetables from the green grocers, the food situation would improve. This, however, did not happen. The seemingly unending years of war brought many hardships to the British people. The shortage of food, notwithstanding how many ration stamps you might have, was certainly not the least of these inconveniences. Practically no meat comparable to the type and grades that had been available to the boys in America could be bought. Other than a few native apples, no fruit was available. Sweets were limited to brown sugar. Occasionally, soda crackers or thin biscuits made with a generous mixture of whole-wheat flour were substituted for the clammy black bread. Sometimes barley bread was available. Barley soup with lamb fat, called by the boys the daily double, was served both at noon and at dinnertime. The story is told of one of the boys, Bob Christy, who bicycled into the Clinton Arms Hotel at Newark for something to eat. The menu as usual was limited. Noting Welsh rarebit as the specialty of the day, he ordered it. But when served, he objected loudly that it was not rabbit at all because he couldn't find any bones!

The beer at the Fox Inn across the road from the monastery gatehouse was rationed and usually black and bitter, but the boys were beginning to accept it. With a little salt it was not bad. The Saracen's Head, the public house in the town of Southwell, was getting attention from those who had a few hours of recreation during the weekends. Usually, you could get a mug or two of light ale at the Saracen's Head bar.

Ray Hileman's talents and love of gardening were beginning to swell with the bursting buds. The shortage of food was also generating in him a high degree of incentive. Not knowing what lay ahead of him, but with the impression that the food situation in England was not good, Ray had included an assortment of garden seeds in his baggage before leaving America. Remembering his conversation about tomatoes that were for sale at the Newton stall in Newark on market day and his developing friendship with the Newtons who operated the greenhouse at the edge of Kelham village, Ray commenced plans for a garden. This was not an easy matter, however, as land was scarce. The cricket field adjoining the monastery formerly used by the theological students, for example, had been plowed up and converted into a public garden by the monks. The entire production was under the supervision of the Food Rationing Board.

At the back of the monastery, Ray located a small plot about thirty feet by fifty feet between the coal and wood house and the chicken shed, both of which were now without coal *or* chickens.[7] With the approval of Father Gregory Ray went to work preparing the ground for raising a few vegetables. From the monastery garden house, he got tools. George Newton would co-operate by permitting Ray to raise tomato, pepper, and celery plants in the greenhouse.

The spring and summer crops from Ray's small private garden were a most welcome contribution to the boys' slender food supply. The tomato crop filled the need for the boys' table and an occasional gift to the monastery kitchen. There were also good crops of lettuce, radishes, green beans, green onions, and similar garden truck. But the real star of the garden show was Ray's celery. To this day Father Gregory, backed by word of the Kelham villagers, will insist the celery grew to a height of four feet. For the first time, George Newton will tell you, the English people of this neighborhood saw celery stalks bleached to a beautiful yellow white. Ray accomplished this amazing feat by wrapping newspapers around the plants and then heaping soil up on the celery as it grew like

7 "Ray Hileman to Ed Holt," May 20, 1943, *E* file.

Jack's beanstalk. The most astounding thing about the celebrated celery to the Britishers was the fact the boys actually ate it raw as it came from the garden. "They just put salt on it," George remarked. Today, after 25 years, there are many gardens in the Kelham community raising bleached celery to be served raw. Of course, it is still used for celery soup and as an ingredient of vegetable soup. The traditional English creamed celery remains a popular dish on the dinner menus of many homes and restaurants. It can be said with certainty that Ray Hileman left his mark as a gardener.

The success of Ray's garden, and particularly the celery crop, is attributable, says George Newton and Father Gregory, to rich black soil lying along the banks of the river Trent and to the fact that when Ray came off the morning tour at midnight, he irrigated his garden with a plentiful supply of water from the monastery's water system, pumped up from the nearby river Trent.

During the late spring Ray was transferred from the job of motorman in a drilling crew to driver of Ole Lo-Go. This change, however, did not interfere with Ray's gardening. He merely did his garden work and watering early in the evening instead of the hours after midnight. But the transfer did open up a new avenue of food supply for Ray and the boys.

This new opportunity almost ended in trouble for Ray and a big demerit for the Americans. As Ole Lo-Go raced about the area of the oil fields, Ray began to notice a great many pheasants strolling leisurely along the hedgerows and in the pastures near enough to the road that they could easily be seen.[8] The temptation was too great. Ray's trigger finger twitched with delight while he drooled at the thought of baked pheasant with sage dressing—a few oysters in the dressing wouldn't be a bad addition. It should be mentioned that Ray

[8] Interview with the George Newtons in England during the summer of 1969.

was not only an expert on a drilling-rig floor, a gardener, and a hunter, but also an amateur cook of no mean ability.

It was not long until Ray had managed to find a shotgun. Later developments proved that it was borrowed from George Newton's father along with a few shells. The pheasants began to come into the boys' kitchen often. All hands were delighted. The bonanza lasted several weeks. The pheasant harvest, of course, fluctuated with Ray's ability to get ammunition. He found a few shells in Newark stores, a few at Southwell, and got a few from home in his holiday food boxes. Some say he got a box or two from London. It was suggested that with forty hungry young men working on the shotgun shell problem, even in war-torn England, it would be at least partially solved.

Ray became close friends of the Newtons, and today they will show you the bent forks and spoon used by Ray when he was teaching the Newtons to fry chicken southern American style and to make cream gravy. One day when Ray came into the monastery after working hours, George Newton was there to meet him. Taking Ray out for a walk around the monastery, he broke the news that the farmers of the area were aroused by the mysterious gunshots they had been hearing at odd hours of the day. Closer investigation had revealed that a man on a big lorry was riding about the country shooting pheasants at random. Newton explained that poaching on another's land in England was a high crime and punishable by severe penalties. Newton understood that police officers would be in the area on the following day to talk to the man on the big truck. That night the shotgun and the shells on hand were returned to Newton, Senior.

Next morning, Ray was too sick to work. He thought it best to stay in bed and try to get over a very sore throat. All the American truck drivers were very busy that day, and one of the D'Arcy laddies was secured to drive Ole Lo-Go. What investigation the deputy made of the man on the big truck was never known, but after a couple of days Ray's cold got much better and he felt able to return to work as Ole Lo-Go's driver. Pheasant, however, suddenly disappeared from the menu of the boys' table, and Ray commenced to give more time to his gardening.

After Ray's days of riding shotgun on Ole Lo-Go were over, he had a memento that would come home with him as a memory of his battle for oil in Sherwood Forest. One day he had met a fine specimen of a red fox in the road, and the skin was cured in George Newton's greenhouse.[9] Later, it was a gift to Ray's wife. Ray's enthusiasm to do something about the food situation for the boys was not diminished. A few weeks later, on May 20, he wrote Ed Holt a personal note saying, "I have two laying hens that lay most every day. Those eggs really help too. I also have a hive of bees, and 18 tame rabbits. They call them hares over here." Careful and studious research by the author has failed to determine how the beehive was acquired.

The spring days seemed to promote a growing companionship between the boys and particularly the young women of Kelham village and the surrounding countryside. Meetings at the Fox Inn for a mug of the black and bitter, as the draft beer was sometimes called, were becoming a regular after-hours' pastime. It was here at the Fox Inn that many of the boys forgot their homesickness and the long delays in mail deliveries. E. E. Edens, perched on a bar stool, had taught the British "Deep in the Heart of Texas," while the British harmonized with "It's a Long Way to Tipperary," and many other war ballads of the day.

Meanwhile, Rosser and the truck drivers were meeting the

9 Interview by author with Father Gregory and George and Doris Newton in 1969.

cargo ships that were arriving with drilling equipment in Cardiff, Liverpool, and Glasgow. And the action in the oil fields was taking on the appearance of a real oil-country boom town.

XI.

"Taking Things Here as a Whole . . . They Are Getting Better"

On Sunday, April 11, word came that equipment was arriving at Liverpool, Cardiff, and Manchester.[1] Later the same day, Rosser and Robbie with one of the K-7 trucks were on the road for Liverpool. They arrived before dark and contacted the freight forwarders, Guest Transportation Company. When the gates opened next morning, Rosser and Robbie were on the dock to pick up two motors and several boxes waiting for them. As morning passed and nothing happened, the long wait became too much for the impatient Rosser and Robbie.

When the dock workers closed down at noon for the lunch period, Rosser and Robbie, already boiling with irritation, started operations of their own. Without regard to dock regulations, they hurriedly rigged up the gin poles. By the time the dock workers started returning from their lunch break, the two motors and the boxes were about loaded. As the men came onto the docks, they were amazed to see Robbie's truck standing on end with the front wheels raised high in the air, and looking much like a wild stallion at bay. None of them had ever seen a self-loading truck equipped with winch and gin-pole lines. High in the seat of the truck, Robbie was in command of the winch, while Rosser was energetically attach-

[1] Rosser's diary, April 11-14, 1943, and interviews and tapes are sources for this information.

ing chains and loading hooks to the equipment that was to be pulled over the lowered tailgate onto the truck.

The astonished dock workers, sure that the truck and driver were in great danger of being pulled over backwards, started yelling in unison, "Hold it! Hold it! Hold it!" Robbie, not aware of the dock workers' unfamiliarity with such loading operations and thinking something was perhaps wrong, shut off the power and called to Rosser to find out what the trouble was.

"Nothin'," yelled Rosser, "these damn Limeys don't know what the hell they're seeing or what they're talking about. Go ahead, winch up the load." Less than two hours had been spent in loading the equipment. Today, Rosser and Robbie insist the job would have taken the dock workers a full eight-hour shift to have loaded the material with their traveling cranes. The operation had been so spectacular that the dock authorities were spellbound. They did not seem to realize there had been a blatant violation of port regulations. A dock foreman rushed up to Rosser and wanted to know where such equipment had come from. Could he and Robbie, with the truck, be employed for work on the docks? Rosser assured him that trucks in America were regularly equipped with winches and that he and Robbie already had all the job they could handle.

Without further delay Robbie, with Rosser at his side, drove the truck off the dock and headed for the gate opening onto the highway. A number of trucks were queued up in front of them and moving very slowly. Guards from the dock master's office were checking bills of lading against the cargo aboard each truck, a delay which was another irritation to the Americans. Robbie, at Rosser's urging, continued driving, whereupon two guards jumped on the side of the truck. Robbie again wanted to know if he should stop. "Hell no," was Rosser's answer, "wipe the sons-of-bitches off!"

Robbie swung close to the gatepost on the side where the

guards were riding. They were forced to jump from the truck. A block away Rosser looked back. A bobby was following on his bicycle. The bicycle was, of course, no match for the truck. The equipment was now well on the road to the Eakring yard for assembling into one of the National 50 rigs. The motors and other equipment were delivered at D'Arcy's Eakring yard before midnight. Despite their violation of the dock's rules, Rosser and Robbie could not help enjoying the beautiful full moon that made driving almost like daytime. On the next day, however, Rosser thought better of his rash actions on the Liverpool docks and returned to Liverpool and as humbly as possible made peace with the dock authorities.

As shipments of new equipment eagerly awaited by Rosser, Robbie, and the other truck drivers arrived during April and May, they were busy picking up parts and equipment almost daily at ports on the west coast from Cardiff to Glasgow. Rosser was finally able to note in his diary on April 18, "Our first rig spudded today. We had quite an audience, the English boys are really anxious to see us perform the miracles that they have heard us talk about."

The boys finished rigging up on Rig Number 2 on April 23 and spudded another well on the same date. The next day Rosser noted, "Rig Number 3 started tour today—we will spud before midnight if we can get enough water." Since Rosser now had three rigs in operation, he felt the project was at last actually under way. The next week, he was able to return the big old-style rig to D'Arcy. It had drilled two wells under the supervision of J. W. Nickle and Horace Hobbs, morning and evening tour drillers. The principal worry concerning field operation now was waiting for the arrival of equipment reordered as a substitute for the equipment lost to the enemy subs. Delay in getting some badly needed truck parts was a big headache.

Don Walker, in the meantime, was busy with administrative problems. The correspondence passing between P. M. Johns, of the Noble Tulsa office, and Walker, by direct mail and cables via Anglo-Iranian's New York office, was piling up faster than Walker could answer it. By reason of the special air-mail pouch privilege of Anglo-Iranian, such business communications were much faster than other mail. The boys' personal letters continued to be exceedingly slow in getting through because of the delays in routine censorship to which all private mail was being subjected.

Procedures for handling and reporting of the payroll of the boys on a bi-monthly basis, accident reports to Tulsa and to the insurance carriers, tax deductions, and other information needed for accounting purposes and for compliance with the many applicable wartime laws, rules, and regulations of both countries were certainly complicated and tedious chores and were dreadfully time consuming. Don was responsible, too, for furnishing D'Arcy and his home office the Drilling Report —the bible of oil men which furnishes at a glance progress made in the drilling operations and trouble areas encountered.

Walker's constant problem which he lived with twenty-four hours a day was the search for food that could be had on the limited rations allotted to the boys. The rationing board located in the town of Southwell had jurisdiction over the rations for the Americans. Green stuff, in particular, such as tomatoes, lettuce, and celery, seemed always to be in short supply. Mutton and barley were standard.

Another responsibility that fell to Walker was administering first aid to the sick and injured.[2] He was charged with the duty of making sure the men received medical attention at Kelham Hall or the American general hospital at Mansfield, located some twenty miles east of their living quarters. Rosser, Walker, and Southwell had been able to make standing arrangements with the hospital staff to furnish medical attention

2 "Rosser to Holt," April 13, 1943, E file.

for any of the boys whose illness or injuries might require the care of a doctor or admission to the hospital.

Noting Walker's heavy responsibilities, Rosser got a note off to Ed Holt in which he quipped that Walker was busy as a cat on a tin roof. He hoped, he said, to get Walker knighted for the wonderful job he was doing. "Besides," he told Holt, "Walker's girl friends would enjoy the prestige of being on 'Sir Donald Walker's' mailing list."[3]

C. A. P. Southwell, Sir William Fraser, J. A. Jameson, and other Anglo-Iranian officials with whom Rosser and Walker came in contact were elated that some petroleum products were now coming through the Mediterranean from the company's 400,000-barrel refinery on the Persian Gulf, at that time the largest refinery in the world. Because a considerable part of the Anglo-Iranian refinery's output was going to the Russians and to allied forces in the Pacific theater, tanker shipments to Britain would be curtailed for some time.

It now seemed clear that the Russians, after two years of defending the long German front, had turned back Hitler's Panzer divisions in their race for the oil fields in the Caspian and Black Sea areas. Anglo-Iranian executives, however, were sure that more of its Abadan refinery output would in the near future be permitted to flow into the European theater to fuel the channel invasion of enemy-held France now being optimistically discussed by many war-weary Britishers.

Back at Kelham, Walker's scramble for food was relieved to some extent under an arrangement with the rationing board which allowed the monks to deliver fresh vegetables from their garden directly to the boys' kitchen. But already the twelve-hour tours being worked by the boys on the skimpy British civilian food rations were beginning to take a serious toll. Rosser reported the loss of twenty-five pounds at the end of the first month. Several of the other boys reported similar losses.

[3] "Rosser to Holt," July 8, 1943, E file.

Bob Christie lost thirty-two pounds in six weeks. Almost all showed the effect of the restrictive diet. Adequate food supplies for the heavy work being done by the Americans were not available.

Mail also continued to be a big problem. During April and May Rosser's diary reflected the situation. Day after day, his entries complained of no mail from home. Ray Hileman had written Ed Holt that he never believed a letter from home could mean so much. Lloyd Noble got the story concerning slow mail deliveries and, characteristically, decided to do something to help. He ordered six subscriptions for *Time* and *Life* magazines in the names of Walker, Rosser, and Sams for delivery at Burgage Manor office. Similarly, six one-year subscriptions to *Colliers Weekly*, six copies of *Esquire*, six copies of *Reader's Digest*, six copies of *Saturday Evening Post*, three one-year subscriptions to *The Oil Weekly*, three one-year subscriptions to *The Petroleum Engineer*, and three one-year subscriptions to the *Oil and Gas Journal*. But best of all was the *Sunday Tulsa World* which they asked for because it carried their favorite comic strips.

Time magazine replied that cargo space allotted to *Time* and *Life* was only sufficient for mailing copies already on their list for England, and they did not think that the situation would soon change, but the request would be put on their waiting list to be considered in the event of any changes in their status. The *Reader's Digest* replied in similar manner.

The *Saturday Evening Post* answered that although they were still sending copies to England, it had been impossible for them to guarantee delivery to countries in the war zone for several months. They would accept the subscription with the understanding that they would do their best to make deliveries, and without liability if deliveries could not be made. Rosser and Walker do not now remember whether all the magazines came through regularly or not, but they do remem-

135

ber the enjoyment the boys had from Joe Palooka and the news from home carried in the *Sunday Tulsa World*. Letters from home, however, continued to be their chief interest.

Food shortages and mail delays did not materially slow the boys' drilling operations. On May 18 Rosser reported on the first month's operations. Seven wells had been completed, casing had been set in two more, and casing would be set on the tenth well on May 19. Only three rigs were working. One rig had been put together with help of parts furnished by D'Arcy from its storage yard. The rig was converted to electric power in view of the long delay in receiving shipment of the Waukesha engines as originally ordered.

Rosser noted that water supply had been quite a problem but was improving. He hoped that two water wells being deepened would adequately supply the water requirements for drilling. He also found no particular difficulty in drilling the wells that was not encountered in similar operations in the States. There was some trouble, he said, with lost circulation in the surface formations. This was more noticeable when the drilling locations were on top of a hill. He wrote that he had experimented with peat moss in the drilling mud, and also had contacted a flax mill to get some shavers or husks off of the flax for experimental purposes.

D'Arcy, he said, had decided to do some wildcatting and one of the American rigs in the very near future would be moved to a location about twenty miles from its present site. Rosser explained he had hoped to keep the rigs together where the interchange of equipment and men would be convenient and, thus, help speed up completions. The British need for oil continued to be so pressing that D'Arcy was desperately commencing the wildcatting operations on new geological and geophysical reports that looked promising. "I sure hope they find something big, because they need oil so bad," Rosser wrote.

Rosser was happy, because he had persuaded D'Arcy to cut the waiting time on cement from seventy-two hours to forty-eight hours. D'Arcy was also agreeable to eliminating drill-stem testing of the casing's shoe, which previously had been the general practice in the field. Rosser did not think the extra time required for testing the casing to see if the cement was holding the surface water was worthwhile, since they were circulating cement to the surface. "Out of all the wells that now have been drilled by us and the company in this field, not one has been found that did not have the water shut off," Rosser noted. He suggested that should water be found to be coming from around the casing shoe on completion, it could be repaired easily before any likelihood of damage from water.

Rosser wrote Holt also that the equipment needed to complete the fourth rig for service would require D'Arcy to shut down one of its rigs in order to furnish a derrick and two 150-h.p. electric A. C. motors. D'Arcy was agreeable to the plan because they were convinced it would speed up completions. Generally, Rosser thought the operations were moving smoothly:

The D'Arcy boys are fully co-operating. They now realize that to keep our completion rate up, the rigs must be serviced promptly and our needs anticipated wherever possible. They are even moving cement and cementing equipment to drilling locations on the same day we start moving our rig. The practice heretofore has been never to cement a well at night. We are cementing them when the pipe gets on bottom, even on Sunday night. We have succeeded in getting the company machine shop to be available on a 24-hour basis for making liners.

The only bad news Rosser had to report was that two men had been hurt. Each had a broken arm. Bob Webster had caught an arm in the clutch on a motor while it was running and sustained a nasty break. A. A. Morton had an arm broken by an endless spinning rope while running casing; only a small bone was cracked, and the doctor had assured him he should be able to return to work within six to eight weeks.

Rosser informed the Tulsa office that since the D'Arcy rig had been returned to the British company after drilling two wells with it, he had taken two men out of the D'Arcy crews to replace the boys who were hurt.[4] The rest of the crews were being used to run casing and to get the fourth rig put together.

Rosser did not miss the opportunity to tell Holt, "We have not been receiving very much mail, but the magazines Mr. Noble ordered have started arriving. All of us are very grateful for his thoughtfulness."

Rosser and the truck drivers were impatient at not receiving the spider gear and other spare parts needed for one of the K-7 trucks. He wrote in his diary on May 10 that he thought he would go to Liverpool tomorrow and try to get the needed parts. Instead, the May 11 diary entry reveals that Rosser went with Rogers of National Supply, and Fraser of Anglo-Iranian to look at two wildcat locations near Nocton and Barkstowe. One of the American rigs would be moved to Nocton and a D'Arcy rig would go to Barkstowe. Rosser understood the wildcats would probably test formations below five thousand feet.

On Wednesday, May 12, Rosser finally visited Liverpool, but did not find the needed truck parts. However, a load of three-and-one-half-inch drill pipe and a box of pump parts had been delivered on the dock, and twenty-nine boxes of equipment were arriving on another boat. Rosser called Eakring to send trucks and by the evening of Saturday, the 15th, the additional equipment was delivered at the Eakring yard.

---

[4] In later years, Sir Philip Southwell proudly pointed out that the apprenticeship of British oil-field workers on the American rigs in 1943 served the Anglo-Iranian Oil Company by furnishing qualified men who carried on the drilling and exploration for the company in Iraq, Iran, Burma, and the East Indies.

Despite the grumbling about the quality and quantity of meals served under British rationing, drilling operations continued satisfactorily. But all was not smooth going in Rosser's world. On Sunday night, May 16, Rosser recorded:

We had a bit of static in our boarding house tonight. Four of the truck boys came in beered up and started raising hell in the kitchen. In fact, Moss boxed one of the stewards in the mouth and knocked out several teeth. I sure hope that they got everything off their chests because we can't stand for very much of this kind of goings on.

Next day Rosser continued:

Don had to take Moss to the Mansfield hospital this morning. His hand is infected and swelled up from the cut he got when he hit the steward in the mouth. Don and I lectured the boys and things have been very quiet and peaceful around here today. The weather is beautiful, but I still don't understand why I can't get any mail from home.

Friday, May 21, Rosser noted, "Today is my darling daughter's seventh birthday. I looked for something in Nottingham to buy and save for her until I go home. I did not find anything. I'd be very happy if I could spend the day with her. I pray she is well and happy on her birthday." On May 22 he was indeed a very happy man. He received his third letter from home on that date and went to sleep the next night re-reading the letter for the third time.

The boys had some more excitement on May 25. The drilling operations of the Americans became the subject of a documentary movie that would be carefully withheld from the public until the censorship on the project was lifted. Southwell told Rosser that the movie cameraman would be in the Eakring field during the day taking pictures of the boys at work. He would like Rosser to be present to help him get some good shots because the cinema was being made for educational purposes and to help teach D'Arcy personnel to do their jobs more efficiently on company rigs. Rosser wrote in his diary of that date, "Don't know whether I'll turn out to be a movie actor or a director before this deal is over."

The boys had completed forty-two wells by June 1. A wildcat location at Nocton was rigging up to start drilling operations. The Anglo-Iranian production department and the D'Arcy men were amazed at the speed of the American well completions, but now they were convinced that such speed was merely a routine reality of the Americans. Well completion records of the Americans when compared to British operations raised questions in the minds of D'Arcy and Anglo-Iranian people as to what was wrong with their drilling operations. There was a great deal of discussion about the drilling practices of the American crews. Finally, Southwell, who was in daily contact with operations in the field, suggested that he bring Rosser to London for a meeting with the production department and Anglo-Iranian officials to discuss time-saving methods of the American crews that might be put into practice by the company employees. The invitation tendered Rosser by Southwell admittedly elevated his youthful ego.

XII.

"St. Peter's Waiting Room"

"I think your boys are trying to live too much by the book."
Rosser nibbled at his unlighted cigar. "I guess it's all right to
have written instructions as to how things are to be done, but
the men should understand they are free to alter their procedures when necessary. Things don't happen the same way
every day because conditions and circumstances are different
at almost every location." Rosser was desperately trying to be
polite and at the same time get his message over to the distinguished group of officers, directors, and other key personnel
of the great Anglo-Iranian Oil Company, Ltd.[1]

The setting was undoubtedly an over-aweing experience for
the young, cocky American. The ornate mahogany table that
stretched the length of the equally ornate board room of the
largest oil company in the world must have looked something
like the polished floor of a bowling alley lane to young Rosser.
The great highbacked, soft, green leather chairs were occupied
by Sir William Fraser, chairman of the board, at the head of
the table; flanking him on either side were Sir George L.
Barstow and Sir Edward Packe, representatives of His Majesty's government. At the far end of the table sat Mr. H. B.
Eves, deputy chairman. Near the center on one side of the

[1] Rosser's meeting with the board of directors of the Anglo-Iranian Oil
Company was told to the author on several occasions.

table sat thirty-one-year-old Rosser, supervisor of the drilling crews of Noble Drilling Corporation and Fain-Porter Drilling Company of the United States that were now operating the four drilling rigs in the oil fields of the British Midlands for the D'Arcy Exploration Company, the wholly owned subsidiary of the Anglo-Iranian Company.

Alongside Rosser was Mr. C. A. P. Southwell, managing director of the D'Arcy Company. On the opposite side of the table sat John T. Cargill, Sir Warren Fisher, N. A. Gass, J. A. Jameson, president of Anglo-Iranian, Sir John B. Lloyd, F. W. Lund, Frank C. Tiarks, Robert I. Watson, G. C. Whigham, and H. T. Kemp, secretary, all members of the board. Slightly back of Rosser and Southwell and along the wall back of the directors sat members of the production department and other employees having to do with exploration and development of both companies.

At one end of the table Rosser was conscious of a young lady pushing the silent keys of a stenotype machine that recorded his every word. He could not fail to be conscious of the deep pile of regal purple carpet in which his feet were planted. The high-domed, ornate ceiling of snow whiteness reflected a soft indirect light over the big room. Rosser remembers a fleeting thought racing through his mind that "St. Peter's waiting room must look something like this."

Southwell had introduced Rosser by explaining he was directing the operations for the American companies with which D'Arcy had contracted to furnish four drilling rigs suitable for development of the oil reserves in the Eakring and Duke's Wood area. Drilling operations of the Americans had commenced late in April and during the short time the Americans had been on the job, Southwell said that forty some odd wells had been completed and put on production. The amazing accomplishments of Rosser and the drilling crews, Southwell said, was an impressive example of know-how and efficiency. He was confident that much could be learned by the Anglo-Iranian and D'Arcy companies by studying the practices and methods of the Americans. He had "therefore urged Mr. Rosser to meet with us for a critical analysis of our company's drilling and development operations in the Eakring and Duke's Wood fields."

Rosser had met Sir William Fraser and Mr. Jameson in the field.[2] Their questions and comments were friendly and encouraging. As the meeting proceeded, Rosser's confidence and assurance grew with his enthusiasm. "Actually, I believe that when your men forget the rules as laid down in the books and do the job at hand in such way as to get the best and fastest results, their time in completing and putting the wells on production will be considerably reduced. For example," Rosser pointed out, "your rules require the drilling crews to change rock bits at specified times and in many cases the cutting speed of the bit is still doing a good job. But your book says change the bit, so the bit is changed. We never change a bit as long as it is satisfactorily making hole. Much time is lost in pulling the drill pipe out of the hole and running it back to change a bit that does not need changing. A driller who knows his business can certainly tell when the bit in the hole is doing a good job or not."

Rosser went on to explain other time-saving devices practiced by the American crews, such as drilling with water when the formations are known and no great pressures are present. Heavy drilling mud is necessary only when you may encounter formation pressures that must be controlled. Heavy mud tends to increase the drilling time, while drilling with water where it is safe to do so, as is the case in the Eakring and Duke's Wood fields, tends to speed up the time of completing the wells. Long waiting periods for cement to set in these fields are unnecessary.

[2] Rosser's diary, entries for July 1, 2, 1943.

"At present," he told the group, "your field men are waiting seventy-two hours before testing cement jobs on the wells they are drilling. We have reduced the waiting time to forty-eight hours and have not yet found a single well in which cement has not shut off the water." Rosser went on to discuss the time that could be saved in skidding the rigs as a unit from one location to the other instead of the usual practice of dismantling the equipment and reassembling it at each new location. Having water and fuel connections at the location by the time the rig is on the drill site, having the cement at the location ahead of the time needed, and several other time-saving practices were suggested by Rosser.

"The most important thing," Rosser insisted, "is to give your men on the job authority to use their judgment as to the methods and practices to be followed. Of course, some mistakes will be made, but the end results will, in my opinion, compensate for any such mistakes."

Rosser grew expansive, and with great pride told the British gentlemen that the man he worked for, Mr. Lloyd Noble, was a highly successful drilling contractor and businessman who believed in encouraging his men to assume responsibilities.[3] "Lloyd Noble," Rosser told them, "holds the view that machinery and drilling equipment are nothing but a pile of iron unless used and operated by competent and efficient men who are capable of exercising judgment and discretion as dictated by circumstances. Mr. Noble repeated this philosophy often," said Rosser, "and was careful to never let his men forget this truth."

The Anglo-Iranian officers and directors seemed impressed. The truth of Rosser's comments was now being confirmed in the field. It was a fact that the American crews were completing and putting on production wells in the Eakring and Duke's

[3] Noble had written to Rosser, Walker, and Sams on June 15, 1943, "We realize that irrespective of how good machinery is, it won't dig wells," *E* file.

Wood area at the average rate of one well per week. The British crews' best time on completing a well for production had been five weeks. In most cases about eight weeks were required by the British crews for completing and putting the well on production.

The result of the frank discussion between Rosser and the company officials was an arrangement whereby one British worker was put on the drilling rigs with each of the American crews, or "teams," as the British preferred to call them. The plan worked well. The British were in most cases anxious to learn and the American crews proved to be extremely co-operative.

In justification of the British drilling crews, it should be pointed out that the rigs they used were designed for much deeper drilling than the producing formations of the Eakring area and consequently were cumbersome and awkward to operate. British drilling crews, except the drillers, were largely made up of inexperienced men, and in some cases were physically not fully capable of the strenuous work required in the oil field either because of age or physical handicaps which had resulted in rejection from military service. Deferment from military service such as was followed in the United States was not the case in England. The desperate situation which Great Britain faced in 1940 had required all able-bodied men to submit themselves for military service. Those possessing skills required for the production of war material were assigned to such occupations by the military authorities.

Common labor as used generally in the oil field was drawn from the labor pool formed in the various districts. The employer made application to the labor pool, and without regard to the nature of the work to be done, he was sent such employees as might be available. In the case of the work being conducted by D'Arcy in the oil fields, the majority of the men were wholly inexperienced in oil-field work. Three years

145

of war food rationing had likewise taken its toll of strength and vitality of a people handicapped by the heavy psychological burden of the war.

Yet another difference between Americans and Britishers was the influence of the ancient guild system that set men apart in the category of their social, economic, and in some cases, religious lives. The tradition of the nature and character of a man's work was largely determined by the guild into which he was born. This system of social and economic levels determined largely the hours of work and the recognition of holidays and holiday periods. Even to this day, with the exception of the nationally recognized bank holidays, the people of Lincolnshire, Nottinghamshire, and Yorkshire arrange their holiday periods on different days of the year so that the holiday of the people of one county does not overlap or interfere with the holiday of another.

In the early 1800's, the guilds had gained the right to bargain with employees on hours of work and the employment of apprentices. Later, Parliament passed the Workmen's Act in 1825 permitting the creation of labor and trade unions. Today, the rights of the workers, both by law and custom of the trade unions, are well recognized, including working conditions and minimum wage laws. The eight-hour working shift for the British drilling teams, as the British called them, and the observances of their many holidays are well-established practices followed by many trade unions during the war years.

The entry of Monday, June 14, in Rosser's diary comments: "Everybody with D'Arcy is having a holiday today. Whitsun Sunday or some such damn thing. They are as bad as the Negroes in Louisiana."

One could speculate that these established rules of the trade union, both by tradition and law, may have been the reason for the all-inclusive authority of the military services in the assignment of those with special skills to certain activities. For example, the engineers of the aircraft industry, although members of the military services, were reassigned to the aircraft plants.

The tempo of the activity in the field was stepped up during June. The Americans now had been able to put together three rigs which had been made possible by borrowing some parts from the D'Arcy Company. With the 13 D'Arcy rigs in service, the American rigs, becoming rigs numbered 14, 15, 16, and 17, would go into service as soon as the mast and the Waukesha engines arrived.

Almost a million barrels of crude oil of exceptionally high paraffin content would be shipped by pipe line and tank cars to small refineries near Liverpool and in the south of Scotland before the year ended. This was only a drop in the bucket compared to the oceans of oil required to satisfy the allied army's ever-increasing hunger for petroleum products. When evaluated in terms of lubricating oil, of which there was a severe shortage and a large imperative demand, the production of the British Midlands at this critical period took on a new and greater significance.

Rosser, in the weekly letter to the home office, quoted a British idiom: "Everything is 'cracking' now." The boys had lost two of their number; one had quit voluntarily, and the other had been sent home as an incorrigible troublemaker. But by the end of June, Ed Holt and Red McCarty, of Noble's Tulsa office, had been able to send over A. L. Long and Lowell Purdy as replacements. Everything would certainly be going their way if it were not for the food situation.

XIII.

To Eat or Not to Eat,
That Was the Question

Through June the Americans maintained the rapid completion of wells that had so amazed the employees of the D'Arcy
Company at the start back in March of 1943. They were
limited to the three rigs they had been able to assemble with
the shipments that had safely reached British west coast ports,
since replacement of the equipment lost to enemy submarines
had not yet arrived. By July 5 the three rigs were credited
with the completion of twenty-five wells. Completions would
go faster, of course, when the fourth National 50 rig could be
put into service. If the two badly needed Waukesha engines
would only arrive, it would help to step up operations.[1] One
of the rigs had been shut down for several hours on two
occasions because of electric power failure. Despite the shortage of power, however, and the shortage of skilled labor, the
electric companies were doing a good job of keeping electric
energy coming to the rigs. The drilling crews, when possible,
were helping the company men with bringing the power lines
to the new locations.

One matter that was bothering Rosser at this period was the
lost time that was beginning to show up on the American rigs

[1] Cartwright Reid advised Edgar Holt on June 17 that the engines were
scheduled for delivery July 15. Rosser by cablegram, June 5, urged a speed up
in shipment of spare parts and equipment, E file.

because of well troubles. In most cases the lost drilling time was due to lost circulation on fishing jobs. Rosser knew from experience that in most such cases the trouble was attributable to a careless act of some member of the drilling crew. One of the rigs had lost time fishing for drill pipe twisted off in the hole. At another rig 1,600 feet of drill pipe had been dropped in the hole. The same rig had lost time fishing for bit cones twisted off in the hole. Still another rig lost time because of stuck pipe. Increasing well troubles were keeping tool pusher Gordon Sams busy and irritated. Rosser had talked to the drillers. They agreed the crew members were not as alert as they had been in the beginning, but the nagging well troubles continued. Between trips to west coast ports with the K-7 and K-8 trucks to pick up shipments now dribbling in and trips to Nottingham, Birmingham, and Carlisle looking for spare parts needed for truck maintenance, Rosser spent as much time as possible in the field. Water, also, was proving to be a problem. Getting a sufficient supply was difficult. Failure to keep water pumps in working condition was causing delay.

The desperate need for rapid well completions was forcefully brought home to Rosser, Walker, and Sams by the sudden announcement of the D'Arcy Company that it had decided to increase the wells in the Eakring and Duke's Wood pools by adopting a drilling pattern of one well to each two and one-half acres instead of the five-acre spacing pattern as originally planned. Things were really up-tight with everyone with whom Rosser and Walker had any contact.

It was obvious to Rosser and the truck drivers who had access to the docks at the south and west coast ports that the fast build-up of military supplies, including armament and ammunition, foretold the massive military operations that were in the making. The great quantities of food supplies, continuously arriving at every available port, were especially noticeable. Ships were arriving faster than they could be unloaded.[2] Dockage space was at a premium. Some ships were waiting as much as thirty-six hours to discharge cargoes. Finding shipments of oil-field equipment and getting their trucks loaded and off the docks was a time-consuming problem for Rosser and his drivers. Rosser never ceased to be thankful for winches and gin-pole rigging on his trucks. Rosser told Walker that the south of England was one big storehouse of military supplies.

During the last week of June, while moving rig number 14 to the Nocton location number 1, the first of the number of outside test wells the American crews were to drill for testing formations below five thousand feet, a British military convoy was encountered. It appeared to Rosser to be an army division fully equipped for combat. The convoy stretched along the narrow curving roads for more than seven miles. The large pieces of army equipment and army trucks temporarily blocked the road.

After considerable talk with a British army captain, who appeared to be in command of the rear part of the convoy, the captain finally demanded to know "what the hell is all this machinery for and what has it got to do with fighting a war?" Rosser's response was that "if this equipment wasn't being used in an important war effort we damn sure wouldn't have brought it all the way from America." The captain's attitude softened. After a bit of gentle urging that made the captain know that Rosser fully appreciated and respected his rank and responsibility, orders were given to the rear guard MPs to conduct the oil-field trucks past the convoy that was then standing by while most of the men were making morning tea over canned heat. As the MPs moved along ahead of the oil-field trucks, the army vehicles were promptly pulled to one side permitting the oil-field equipment to move along the seven miles of the narrow roadway. As the oil-field trucks

[2] Recorded in Rosser's diary and recalled by him in taped interviews.

Pumping oil well and cattle share same pasture in Eakring oil field, Nottinghamshire. *Courtesy Milton Adcock.*

Pumping unit in Duke's Wood, Eakring. *Courtesy British Petroleum Company.*

Dewaxing operation at Eakring. *Courtesy British Petroleum Company.*

"Ole Logo."

Rosser and Sams in the field.

Some of the boys out for a morning spin. From left to right: Joe
Webster, Glenny Gates, Al Morton, Spanky Hemphill, Chris
Watson (in bathrobe), and Joe Barker.

Afternoon tea.

A. J. May's crew.

Rosser and Brother James.

The J. W. Nickle crew.

American oil-field workers before leaving New York for England. Left to right: *bottom row*, Phillip Albritton, C. M. Reidinger, J. L. Waits, Herman Douthit, H. G. Hobbs, G. G. DeArman, J. E. Harding, Elray Davis, J. S. Barker; *second row*, R. M. Christie, L. P. Robinson, F. E. Moss, Carl Norberg, Clarence Sikes, De Talt Havely, C. E. Olvey, Dewey Aycock; *third row*, K. F. Johnston, E. H. Boucher, J. W. Nickle, G. E. Griffin, Virgil Latham, C. F. Johnson, A. F. Webster, Jr., A. T. Webster, E. G. Gates, L. M. Oaks, G. C. Watson; *top row*, L. B. McGill, J. H. McIlwain, Jr., E. H. Hemphill, A. A. Morton, L. V. Dugger, Ray A. Hileman, Wm. M. Burnett, E. E. Edens, Ray F. Miller, W. W. Walden, A. J. May, Gordon Sams, P. G. Allen, Red McCarty.

passed the army convoy, Rosser noted that it appeared to be a full division of the British infantry completely equipped for combat from field kitchens to truck after truck of ammunition and hundreds of gasoline and oil drums. Obviously, the British were moving men and supplies into a staging area for action anticipated in the immediate future.

Rosser now remembers that after unloading the drilling equipment at the new location on the return trip to Kelham village later the same day, he noted the soldiers accompanying the convoy were now along the roadside preparing their afternoon tea. He observed to Robbie Robinson, who was driving the truck they were in, that the British custom of morning and afternoon teas was doing more to stop the British army than Hitler and his generals had been able to do. Robbie's answer was "that Britain's teas were damn sure no more important to the British than the coffee breaks were to the Americans."

This comment, however, did not seem to impress Rosser who a few days later found some of the drilling crew members making British tea on the job. Rosser's diary reports that he had thrown a "hissy," stomped the pots into the ground, and in forceful language told the crew members that he doubted if any of them had ever before had a cup of tea. In fact, he doubted if any of them could spell it, much less have a compelling taste for the damn stuff. During this period, Rosser was spending considerable time at the drilling rigs and on the road with the trucks, doing everything possible to maintain a high standard of operations.

In the meantime, Walker was having his troubles. Finding the type and quantity of food required to keep fit and healthy forty-four young Americans engaged at hard labor twelve hours a day was getting to be a bigger problem every day.[3]

[3] Mrs. Ann Soloman, Southwell, England, wrote to Don Walker May 19, 1944: "I certainly remember you well. You would go into the Food Office at Southwell quite often to get ration tickets for your men. I worked there and you

Red meat and fresh fruit that had always been important items in the boys' diets were suddenly not available to them. The kind of food and in the quantities required were just not to be had in the English markets. Walker continued to insist that the fishing jobs and the other well troubles the boys were having on the rigs resulted from inadequate food supplies. The field trouble on the rigs, Rosser finally had to admit, was indeed a matter of food.

Walker also said that he recognized a sudden irritability in some of the boys that he had not noticed before. Some of them were doing considerable off-the-job drinking at the Fox and other nearby places where hard liquor could be had.[4] The British whiskey rations issued to the Americans were pooled and available to them in the recreation quarters during their off hours. The boys were, of course, well aware of the strict rules prohibiting drunkenness at any time. The use of liquor in any form during working hours was not allowed. Rosser, Sams, and the drillers had made it clear to all the crews these rules would be rigidly enforced. In retrospect, it is a significant fact that during the time the American crews were in Britain, no crew member or truck driver or helper was under the influence of alcohol while on the job.

Rosser was satisfied that the lack of proper food was having a demoralizing effect on the boys, both on and off the job. The problem was not a matter of rations, or food tickets as the British called them, but an actual shortage of food supplies in the markets. In particular, the green grocers had little to offer. The arrangement with the monks to furnish some fresh vegetables from their public garden was helpful, but not

---

had my sympathy in those days for it must have been quite a headache after leaving the U.S. to find such difficulties in feeding hungry men," *E* file.

4 This was confirmed by the author in an interview in England in 1969 with Mrs. George Lloyd, the former Binty Hayward who had been employed at the pub in 1943.

enough to fill the appetite gap of the forty-four hungry oil-field workers. Ray Hileman, who had ceased his practice of shooting pheasants, was now busy tending a private garden and raising a few chickens during his off hours. But his good efforts did not, of course, satisfy the enormous appetite of the rough-necks. Eggs were seldom available. Meat, such as beef, bacon, or ham in sufficient quantities, was out of the question and the old standby, mutton stew, was getting to be an ugly word at Kelham Hall.

The crisis came one morning early in July. The steward blandly announced there was no food for breakfast. He brought on a dish of cold Brussels sprouts left over from the night before. Rosser and Walker insisted the boys make break-fast from the cold and greasy Brussels sprouts. Walker had, in fact, known of the crisis the day before, but since the ration stamps on hand were not valid until the next day and so little food was to be had even with ration tickets, there was not much he could do. He had on several occasions been permitted to receive food against future ration tickets. He would try to do so again today, but he had been warned by the grocers he could not make a practice of over-buying the boys' rations.

A few days before, the boys at this critical period had been given what all would long remember as a great treat. An American friend of one of the boys stationed at a nearby air base had mysteriously shown up with a large sack filled with steaks for everyone.[5] Where and how the friend had come by the steaks is not known but somehow it turned out to be a steak for every one. Rosser and Walker were impressed and re-marked on the great occasion in their diary and letters. The incident put Rosser and Walker to thinking about finding an avenue to the American army food depots.

While on a quick trip to Nottingham, Rosser had met and visited with Major Jack Scott whom he had known in the

[5] Rosser's diary, June 5, 1943.

states. Major Scott had agreed with Rosser that there should be some way to get army rations for the American group since they were engaged in what should be classified as essential war work. He offered to help in any way he could. The idea took firm root in Rosser's mind, and Walker encouraged the idea by the observation that certainly nothing could be lost by trying. Rosser and Walker agreed the situation was critical and decided the matter should be put squarely up to the American Embassy. On Wednesday, June 23, Rosser wrote in his diary, "I am going to London tomorrow and see the American Embassy about food for our men."

On June 24 Rosser checked into London's Savoy Hotel and after a few drinks he records he had a "fairly good dinner." The next day he commenced his campaign for food by contacting George Abbey, information officer in the Embassy, and also Major Brown, a military attaché.[6] They sympathetically heard his story but did not offer any constructive suggestions. Finally, Major Brown told Rosser he should see Colonel Spencer at the headquarters of Services of Supply, European Theater of Operations, United States Army, located in a suburban area about twenty miles out of London. The rest of the day was spent in the London tubes, on buses, and finally a long walk—only to suffer the disappointment of finding that Colonel Spencer was not in his office and no one seemed to know when he would return.

Since Rosser had at last found the Services of Supply headquarters, he was determined to tell his troubles to someone in authority. Without knowing just why, he walked into the office of a Major Nixon. There he discussed the purposes of his trip to London. Major Nixon listened with apparent interest but said he did not know of any procedure by which army food rations could be issued to civilians. He did suggest, however, that Rosser see Colonel C. G. Irish of the Petroleum

6 *Ibid.*, June 24, 25, 26.

Office, Services of Supply with offices at the American Embassy. Colonel Irish was an army adviser to David Shepard, United States petroleum attaché at the Embassy.

The following day Rosser was able to see Colonel Irish and tell him of the pressing need for food if the Americans were to continue the important project then underway. Lieutenant Commander Howe, USN, was called into the conference. He and Lieutenant Commander F. W. Michaux were naval advisers to Shepard. Rosser was invited to lunch at the Embassy by Lieutenant Commander Howe. He now remembers the guilty conscience that mocked him while he put away generous servings of the well-prepared food. Following the lunch, Irish and Howe accompanied him into the offices of Shepard.

The American Embassy at number 1, Grosvenor Square, London, Rosser says, will always be hallowed ground to him. For it was here he struck gold by obtaining the whole-hearted and enthusiastic support of Dave Shepard in his effort to get additional food for the American drilling crews and thus saved the war project that some Britishers today will tell you was one of the important incidents of World War II and, they never forget to say, one of the best kept secrets of the entire war.

The meeting of the four men resulted in action. Shepard pointed out there was only one man who could give the Americans the assistance Rosser sought. That man was Major General John C. H. [Court House] Lee in command of all supply services for the U.S. Army in the European Theater of Operations. Shepard got Lee on the telephone in a matter of minutes. The call culminated in a request by Lee that a written application be filed with his office for the issuance of army food rations for the forty-four American oil-field workers. Before leaving the Embassy office on that Saturday afternoon, Rosser signed the formal application.

At the suggestion of Shepard and for the purpose of supporting the written application that General Lee had requested, it was agreed that Colonel Irish and Commander Michaux should make a trip to Kelham to investigate the drilling operations in which the American oil-field workers were engaged as well as the food situation at the Kelham hostel. A strong report by Colonel Irish and Commander Michaux based on their personal observations, Shepard felt, would emphasize and back up the written request he would immediately send to General Lee.

On Tuesday, June 29, David Shepard supplemented Rosser's application to General Lee with a strong letter of recommendation.[7] He informed the general that the American group consisting of forty-four men worked twelve-hour shifts seven days a week while the British workmen with whom they were associated worked an eight-hour day for six days a week, but both groups received the same food rations per man. He pointed out the excellent record of productivity accomplished by the Americans, then warned that the rate was declining because of their restricted diets. Shepard reminded the general how important authorities in Washington looked upon the production of oil in Britain, that the Americans engaged in that activity were making a fine contribution to the war effort. Then he requested that authorization be granted to requisition supplementary supplies of meat, sugar, and canned fruit from the army depots. He added that it would be deemed a great courtesy if the men were granted the privilege of obtaining cigarettes from the post exchange of the American military hospital located in their vicinity.

Colonel Irish called Rosser on the telephone at Kelham Hall on Tuesday, July 6, to arrange for the visit he and Commander Michaux had promised to make. Rosser was in luck because a Mr. Butterworth from the PAW office in Washington was

7 "D. A. Shepard, petroleum attaché, American Embassy, to Major General John C. Lee, commanding Services of Supply, European theater, June 29, 1943 (copy), *E* file.

also making an official visit to the oil fields and Kelham hostel at the same time. On the following day the Colonel and the Commander arrived on an early train from London. With the assistance of Southwell, the visitors toured the field and sampled the food available at Kelham hostel. The two investigators spent the night at Kelham Hall and were put on the train for London Friday afternoon. Rosser was assured by both Irish and Michaux that they would strongly recommend that army food rations be issued for what was quite obviously an important war operation. Butterworth observed that on his return to Washington he would report to the PAW on the drilling operations and how the short food situation was affecting the American drilling crews. Rosser, Southwell, and Walker were pleased with the assurance of co-operation tendered by the visitors. If the Quartermaster Corps of the Armed Service Forces would co-operate, their food crisis should vanish.[8]

[8] The United States War Department was reorganized March 9, 1942, and the Army Service Forces, the Army Ground Forces, and the Army Air Forces came into being as the three major commands.

The Army Service Forces was a hodgepodge of services, mainly procurement and supply, headed by General Brehon Somervell in Washington. General George C. Marshall selected John C. H. Lee, lieutenant general, to be in charge of the Army Service Forces in the European Theater of Operations which came into being June 8, 1942. Lee's chief quartermaster, responsible for all supply depots in the European theater was Robert M. Littlejohn, brigadier general. Both were career army men. Lee, a graduate of West Point in 1909, had pursued his career as an engineer. Littlejohn, a graduate in 1912, served with the cavalry until World War I when he transferred to the Quartermaster Corps where he had written or approved all army supply procedures.

XIV.

We Eat

The petroleum board's regular meeting was every other Tuesday in London. Dave Shepard telephoned Rosser Sunday evening suggesting he should come into London Monday and remain over to attend the board meeting on the next day, July 13. Shepard did not elaborate on what he had in mind, but Rosser assured him he would be there some time on Monday.[1] Shepard was pleased and said his office would make reservations for him at the Claridge Hotel in Brook Street.

Afterwards, Rosser confided to Walker that there was something in the wind. "I couldn't tell from Shepard's conversation whether General Lee has decided to feed us or starve us, but he sure gave me the feeling that he had some news for us. Since he didn't spill it on the telephone, I'm afraid it is not good."

Walker was more optimistic. He didn't see how General Lee and the petroleum board could expect an all-out twelve-hour day, seven-days-a-week effort from any group of men unless they had the proper food. Walker pointed out the American crews, truck drivers, and helpers were working about 50 per cent more man hours per week than the British crews and British roughnecks. Part of this was the result of the man

1 The sequence of events herein related was obtained by the author through taped interviews with Rosser whose memory for names and dates was refreshed by his diary, July 10–14, 1943.

power shortage of British labor and, part by British oil-field workers' adherence to precedents of the guilds that had been a part of the British labor system through the centuries running back to medieval times.

Nevertheless, Rosser could not overcome the apprehension he felt. Walker persuaded him to attend the Sunday morning church services with him at St. Willfrid's, the small fourteenth-century Anglican church standing about a hundred yards back of the monastery on the banks of the river Trent. But the services did not seem to help his low feeling that persisted. Too, it had been almost a month since he had received a letter from home. Rosser always seemed to have a conscious anxiety about his wife and child during long intervals between letters.

On Monday morning, the 12th of July, after an early breakfast of rough oatmeal, brown sugar and powdered milk, barley toast, and black coffee, Rosser was off to Grantham where he arranged to leave the well-used Plymouth. This was a car that once had belonged to a member of Parliament but had been sold to the D'Arcy Company because gasoline was available only for those engaged in essential war work. At Grantham Rosser boarded the Flying Scotsman, Great Britain's crack train between Edinburgh and London. His diary entry of that date was: "The Flying Scotsman broke down, so I did not get to London until 9:00 P.M."

Arriving in London, Rosser was surprised to find that he was registered in a large and very elaborate suite. He knew that the Claridge was one of the best of London's luxury hotels, but he had never before in all his thirty-one years seen anything quite like Claridge's presidential suite. Sad to say, Rosser was to find on the following morning that he had been put in the suite because the reservation had been made by the American Embassy, and the hotel clerk took for granted the reservation was for some American VIP. As soon as the mistake was discovered on the following day, Rosser promptly moved to a single room in the attic section of the hotel. Afterwards, he discovered Butterworth was also registered in the hotel and had left a call for him. Butterworth wanted to come by for a visit and bring three members of Congress who were then in London on some sort of junket. Fortunately, Rosser had brought along a couple of bottles of bourbon and a bottle of Scotch and was sufficiently fortified to entertain the group with a few drinks.

After his guests had left, Rosser went to bed but was unable to shake the feeling of anxiety that had nagged him all day. The long hours on the delayed train trip had not helped. Finally, Rosser took a stiff drink and told himself that whatever might happen with respect to the food problem at Kelham, he had done all he could at this time to remedy the situation. With this self-assurance he at last went to sleep.

Soon after midnight, Rosser was awakened by the wailing of the air raid warning sirens. They brought him out of bed and to a raised window to see what was going on. The Germans seemed to be really unloading on London. The sound of exploding bombs came from all directions. He had experienced two raids before in London, but they had not been so concentrated on the center section of the city as this one. Tonight, it seemed to him the Claridge Hotel must be very near the center of the Luftwaffe's target. Rosser's fatalistic attitude toward bombing raids prompted him to believe that he was as well off in the hotel as in any other place. He had a strong feeling that the air raid shelter, even the London tubes being used for that purpose, were more hazardous in the event of a direct hit than a building such as the Claridge. Rosser closed the window and returned to bed, but did not remember hearing the all-clear signal. The drink and the aspirin had done their work.

The following day Rosser read that 120 Londoners had been killed in one of the city's large public underground bomb shelters. Many of them were trampled to death by the stampede of people attempting to get into or out of the shelter. The raid proved to be one of the heaviest over London since the summer of 1941.

At the petroleum board meeting that morning he listened through the discussions of the coming demands of the military services for petroleum and the quantities that probably could be furnished from currently available sources to the European theater. It was most pleasing to hear the report on tanker shipments now coming through the Mediterranean. The successful defense against the submarine menace in the North Atlantic by destroyer escorts and their newly devised depth charges was most interesting. He did not fail to note that petroleum was the basic ingredient of the TNT depth charges that were sinking German submarines. He felt honored and privileged to be permitted to listen in on the board's meeting.

The big news to him, however, came at about eleven o'clock when members of the board took a short break. Rosser remembers noticing Geoffrey Lloyd, Britain's secretary of petroleum, and several other British gentlemen were being served tea. During the short break Dave Shepard and Colonel Irish hurried Rosser into a nearby cloakroom. With a resounding slap on the back, Colonel Irish exploded: "The Old Man came through like the soldier he is!" Shepard and Irish then hurriedly explained that General Lee had on Friday, July 9, sent a written order to Brigadier General Robert M. Littlejohn, chief of the Quartermaster Services, the commander in charge of food services for all military units stationed in England. General Lee's order had directed Littlejohn to arrange for the issuance of adequate army rations for Rosser's oil-field workers billeted at Kelham Hall.[2] The hang-up thus far was that General Littlejohn had failed to take action.

[2] The *E* file contains a copy of a letter dated July 13, 1943, from General Lee to Sir William Fraser in reference to the order issued to enable "the drillers

Reviewing the incident in later years, Shepard referred to the conference between Irish, Rosser, and himself as the cloakroom conspiracy, or intrigue, in the Embassy.[3] Irish said he had a strong feeling that General Littlejohn was not in harmony with General Lee's order. The action contemplated by the order was definitely contrary to military precedent, policy, and tradition. It was Shepard's idea that since Irish had already discussed the plight of the boys at Kelham after the inspection trip that he and Lieutenant Commander Michaux had made, Colonel Irish should make an effort to persuade General Lee to telephone Littlejohn to find out if the order to arrange for Rosser's men to have military rations had been put into effect, and if not, why action was being held up.

Shepard reasoned that the report of Colonel Irish and Lieutenant Commander Michaux concerning the food situation at Kelham Hall should have by this time reached General Lee and a copy of the report would probably have also reached General Littlejohn. An inquiry by General Lee would be necessary to get the quick favorable action they were all hoping for. Irish agreed that he would attempt to get General Lee's attention immediately following the adjournment of the petroleum board meeting. Shepard calculated adjournment would be approximately 12:30 to 1:00 P.M. Both Irish and Shepard told Rosser they thought it would be well if he were in General Littlejohn's office when the telephone call came through in order that he might answer any questions Littlejohn would have regarding details necessary to carry out General Lee's order.

The cloakroom conference closed with the understanding

---

to continue to perform the superior work which they have been doing in connection with the war work." The general and the chairman of the Anglo-Iranian Oil Company evidently had conferred about the need for supplementary rations on July 10.

[3] These terms were used by Shepard in a telephone conversation with the author about the incident in 1969.

that Rosser would proceed immediately to General Littlejohn's office and make every effort to let the general know he
was there and available for any questions or additional information Littlejohn might desire or need to start the ball rolling.
But above all, Shepard and Irish insisted that Rosser should
emphasize to him the great importance of the food rations as
they related to the continued swift completion of the producing wells in the little-known oil fields of Duke's Wood and
Eakring areas and the need to make them available without
delay.

Rosser reasoned that the campaign to get action from Littlejohn's office was up to him. With a feeling which he later
described as perhaps similar to that of a young GI rushing into
battle for the first time, he resolutely made his way to 47
Grovesnor Square, an address that had now taken on the high
distinction of the headquarters of the Supreme Commander
of the Allied Expeditionary Forces. Rosser was wearing on this
occasion his field jacket, boots, and the insignia of western
American provincialism—the white Stetson hat which seemed
to build up as well as sustain his confidence and determination.
Nonetheless, he was laboring under a nervousness that he
could only lay aside by continually reminding himself that
the boys were engaged in a project that was essential to allied
success.

Had not Noble and Porter both said that they deemed the
boys in the English oil fields as necessary a part of the forces
being thrown against the enemy as the men in uniform now
awaiting action at the military bases throughout Britain and
other parts of the world? Rosser realized that so far as he was
concerned, his own private D-Day had arrived. Success or
failure of the entire project of the development of the oil
fields depended squarely upon getting quick action from
Littlejohn.

It was with this feeling of win or lose all that he entered the

171

large outer office through the big double doors over which the words *Supreme Headquarters, Allied Expeditionary Force* were lettered. Dozens of secretaries seemed to be clattering efficiently away at typewriters amid the bustle of an extremely busy office punctuated with what seemed to Rosser to be the constant and insistent ringing of telephones.

With a tight grip on his courage, Rosser approached one of the well-groomed young ladies in the uniform of a WAC nearest the double-doored entryway and asked her where he could find General Littlejohn's office. Without pausing to look up from her work at the typewriter, the lady pointed to a door at the end of the large waiting room and told Rosser it would be necessary to see the receptionist in that room. On getting the attention of the receptionist who, he surmised, was in charge of General Littlejohn's waiting room, Rosser told her with his usual directness that he was there to see the general.

This elicited from her the stern request that he write his name, address, and the nature of his business on a specially prepared slip of paper she handed to him. He noticed the printing at the top of the paper, "Brigadier General Littlejohn, USA Commanding Services of Supply, SHAEF," which Rosser was to learn stood for Supreme Headquarters, Allied Expeditionary Forces. Without attempting to comply with the young lady's request to answer the questions indicated on the paper, Rosser handed it back to her with the explanation that he was there to see the general about a very important matter. In fact, he explained, that if his visit to see General Littlejohn wasn't important, he would certainly not be bothering her or the general. With a patronizing smile, she explained that Rosser would have to indicate the information required by the paper before she could find out if or when General Littlejohn would be available for an appointment.

Although Rosser's experience was extremely limited in surroundings such as he now found himself, he was one of those persons who possess a natural gift of judging the abilities and sincerity of people at his first contact with them. In this case, Rosser had the distinct feeling that his efforts to be in General Littlejohn's office, as planned by Shepard and Irish, when Lee's telephone call came through were at an end unless he took matters in his own hands. Time was important. So with the courage and brashness of youth, he strode past the receptionist, through the big office, and into the adjoining office in the direction that he surmised might lead to the whereabouts of the general.

But the office in which he found himself appeared to be occupied by a captain and two officers wearing the insignia of lieutenant colonels. Without hazarding further delay or complications, he ignored the officers and authoritatively pushed ahead into another office that he found to be the province of a full colonel who at the moment was busily dictating to a secretary on the opposite side of his desk. It was here that Rosser saw what he was looking for, on one of the doors opening out of the colonel's office was the magic name, Brigadier General Littlejohn, USA. Upon seeing this, Rosser turned toward the door with the same determination that a swift-running halfback would use to dash for an opening in his forward line on second down and ten yards to go.

Both the colonel and his secretary were on their feet, trailing behind him and loudly demanding that he wait a minute. They were telling him he could not go that way. By this time the lieutenant colonels and the captain were racing in pursuit with the seeming intent to stop him at all cost. But before such action was taken, Rosser had swung open the door to confront General Littlejohn, who appeared to be busy poring over a map spread out on the table before him. His florid complexion and short crew cut made him look very much like Big Boy Rhyne, Noble Drilling Corporation's 350-pound giant superintendent of the Gulf Coast district.

173

The general, seeing his private area invaded by a stranger without advance notice, demonstrated surprising agility in springing to his feet. His face turned to a livid red and with flashing eyes, he loudly demanded of Rosser, "Who the hell are you?" and in a voice that Rosser thought was almost hysterical, screamed, "And what the hell do you want?"

Again, drawing on his intuitive appraisal of the situation, Rosser recognized that perhaps appeasement at the moment was the best policy. He started to explain the purpose of his visit, but by this time General Littlejohn had shouted a command at Rosser to, "Get out of my office, and do it damn quick, or I'll have you thrown out!"

Rosser's control was shattered by the general's last volley, and instinctively he returned the fire by loudly assuring the general: "I'll be glad to get out of your damned office when I find out why you are holding up the order you now have on your desk from General Lee." Rosser bore down on the word "order" with considerable emphasis at the same time praying that the telephone would ring. He had the fleeting thought that maybe Shepard and Irish had failed to get General Lee to make the call. After all, he might have to retreat with considerable embarrassment from his assault on Littlejohn's office. But for the moment, he would stand by his guns and, if possible, tell Littlejohn, whether he wanted to hear it or not, that a few days before, General Lee had sent him an order to issue American army food rations to a group of American men under his supervision now engaged in a highly important and necessary war project being carried out in Britain and one that meant a great deal to the military services.

By this time Littlejohn appeared to understand that Rosser was talking about some sort of communication from General Lee's office; at least the reference to General Lee's order brought a meditative pause in Littlejohn's volley of profanity, that flowed out of him as easily as any oil-field worker whom

Rosser had ever known. When the general shouted and gave a long ring of his buzzer, the colonel from the next office dashed into the room. Littlejohn asked if he knew anything about orders for food to be given to a group of American civilian workers engaged in some war project in England.

The colonel responded, "There is a letter in your desk basket from General Lee that might be related to the matter this bird is talking about."

"Well, get it!" shouted the general, "and let's see what the hell this wild bastard is talking about."

With considerable nervous pawing through the pile of letters on the general's desk, the colonel pulled a letter from the heap and handed it to Littlejohn. The general fell back into his chair and raised his feet to a comfortable height on the table before him in what Rosser hoped was a relaxing position, after which the general studied the letter long and carefully. At least Rosser had time to utter another conscious but silent prayer that the telephone would ring.

Finally, Littlejohn handed the letter to the colonel and, turning his swivel chair to face Rosser directly, snorted, "Where the hell is this operation located you're talking about?"

At that question, one of the majors, who had entered the general's office, grabbed a long schoolroom pointer from the corner of the room and rushed to the large scale wall map that he pulled down on the wall from a set of rollers, like a window shade, and assumed a stance indicating that he was ready to point out the location as soon as Rosser gave him the information. The general appeared to be examining the wall map. But the lull in hostilities was not to last long.

Rosser now felt he was on the winning side and was determined to keep the pressure on the enemy. With some emphasis, he told the major, "Put away your damn stick because I'm not going to tell the general or anyone else where this operation is located. It is an important war effort. It is important that its whereabouts be kept secret. I would not tell you where these men are working any more than you would tell me where some military operation that you might be directing was located!"

At this rebuff from Rosser, the general again called up the salty language of his cavalry career. This seemed to relieve his feelings so that he could once again get control of his mental and verbal processes. He now suggested that Rosser retire to the outer office and dictate a letter telling him the number of men who were to have supplemental food rations from the army, as requested by General Lee, and the additional calories required by each man for the type of labor being performed.

Rosser sensed that the general's suggestion was driving him into a cul-de-sac, but at this point he was not to be brushed off in a blind alley. "Now, general, you know as well as I do that I am not capable of writing you any such letter. All I want is for our men to get hold of some of these hams, bacon, and mountains of canned fruits and vegetables that we see piling up on the docks in every port on the west and south coasts of England. We would be happy if we could just get the food that is being wasted in the trash cans of the army and air force bases all over England."

At this statement, the general took off without any apparent effort and, as Rosser now says, he rose straight up towards the ceiling some two or three feet, but succeeded in coming down with another volley of profanity that seemed to have been held back for this special occasion.

Rosser had now decided to weather the storm in silence as long as possible. Surprisingly, his strategy seemed to have some effect on the general for he again yelled for the colonel, whom Rosser thought was perhaps one of the general's aides, and who again rushed in with the letter seemingly sensing what the general wanted. Another long pause ensued while the general studied General Lee's order. Eventually, in a somewhat relieved atmosphere, the general called for some sort of handbook, which Rosser assumed contained information about calories and the care and feeding of men engaged in strenuous work.

It was at this moment that the general's telephone started ringing. Rosser maintains to this day that he had never before or since heard sweeter music than the jangle of the telephone on General Littlejohn's desk. He knew it meant that General Lee was on the other end of the line. The telephone conversation of Littlejohn confirmed that the telephone call was, indeed, the one Rosser had been praying for. General Littlejohn's concentration at the telephone gave Rosser his first opportunity to mop away the perspiration that was seeping down his face and into the neck of his field jacket.

Rosser was hearing himself graphically described to General Lee in a most unflattering manner, but the important thing that came out of Littlejohn's conversation was the happy and meaningful statement:

Well, general, we don't know exactly how we can do it, but we will arrange some way to carry out your order that will be satisfactory to these men and the men in charge of our newest supply depot. Up to now, I have been unable to get this fellow Rosser to tell me where his operations are or any of the details. What you are telling me is the most information I have had concerning the Embassy's wishes relating to this operation.

Rosser was brought back to reality by the assurance Littlejohn was giving General Lee that the matter would be taken care of today. The victory seemed to be so complete that Rosser was already feeling a wave of sincere sympathy for the general who had crumbled before his very eyes.

Without further questions or statements, the general then snatched the letter from the table and strode out into the big room where the typewriters were still busily clacking away.

With nothing more than the urge of curiosity, Rosser followed along. The general was dictating a memorandum concerning rations for civilian oil drillers directed to the Commanding Officer, USA, Depot G-20, and, as Rosser got it over the noise of the typewriters and the buzz of conversation, "that pursuant to instructions of the commanding general, it is desired that you sell for cash to Mr. E. P. Rosser, Kelham Hostel, Kelham, Notts., rations to the value of the difference between American and British rations for those American civilians employed as oil drillers here in the U. K. at the present time."

At this point Rosser felt the bravado draining out of him and sought a nearby chair. Finally, the secretary who was taking the general's dictation on the typewriter finished with a flourish, reached for an envelope, and asked the general if a copy of his memorandum should be sent to Mr. Rosser.

The question was an unexpected shot in the arm for the general. With a bellow that could be heard throughout the big room and that reverberated through the surrounding offices, General Littlejohn, with the fervor of a crusading evangelist, screamed, "HELL, NO, GIVE IT TO THE SON OF A BITCH, NOW!" Pursuant to these forceful directions, the secretary smilingly but slowly handed Rosser a copy of the historic memorandum signed by the general, the provisions of which would mean so much to health and morale of the Americans.[4]

On the subject "Rations for Civilian Oil-Drillers," the memorandum was addressed to the commanding officer of the army depot nearest Kelham, "pursuant to the instructions of the commanding general." It provided for rations to forty-three individuals, effective immediately and extending to December 31, and to be paid for in cash at the rate of twenty-

4 "Littlejohn to Commanding Officer, U.S.A. Depot G-20," July 13, 1943, *E* file.

seven cents each, the computed difference in money value of British and American rations.

For the moment, Rosser had no other objective in life but to get back to the American Embassy to replay the Rosser-Littlejohn conflict for the benefit of Dave Shepard and Colonel Irish. With a copy of the order safely tucked in his hip pocket, he told General Littlejohn and the secretary, "I promise you that my coattails will not touch my rear end until I get back to Kelham and start feeding our starving group." Rosser, realizing that it would be necessary to deal with General Littlejohn again in December for an extension of the privilege to continue purchase of food from the army depot, and also aware that the commanding officers at the army depot where the food purchases were to be made were under the direct command of General Littlejohn, was anxious to terminate the confrontation with Littlejohn on a peaceful note, if possible. Taking both courage and pride in hand, he again walked into Littlejohn's office. Now the doors were open and no one seemed disposed to place any obstacles between him and the general. He found him standing in the middle of his office, studying a copy of the letter that he had just signed concerning procedures for the purchase of the coveted food. With as much humility as he could force on himself at the moment, Rosser told the general he wanted to apologize for all the disturbance he had caused in the office and for any insulting remarks he, in the heat of argument, had made. He hoped the general would understand that it was a mere matter of enthusiasm in trying to get food for his men and would not take what Rosser had said personally. Rosser assured the general that anything that had been said to him (Rosser) he had probably deserved, and as far as he was concerned, the matter was entirely forgotten. Rosser put out his hand for a friendly, parting handshake. Rosser now says that the general slowly put out his hand which

Rosser eagerly grabbed but was somewhat taken aback, because it felt as though he were picking up a dead fish's tail. The general's hand was soft, pudgy, cold, and unresponsive, which was sufficient encouragement for Rosser to drop the parting apology and rush out of the office. The general's attitude now meant little to him, since he had in his possession the precious order directed to the army depot. He felt, however, that he must first thank Shepard and Colonel Irish, and he must, of course, call Walker.

When at last Rosser got Walker on the telephone, Walker's comment was a memorable one. "That's what I call literally bringing home the bacon!" Walker's voice was thick with emotion. In later years Dave Shepard, now a retired former member of the board of directors of Standard Oil Company of New Jersey, characterized Rosser's battles for American army food rations to supplement food being received by his crew as a fight in which Rosser "whipped the U.S. army and almost literally took the food for his men out of a supply depot near the oil fields in the British Midlands."[5]

Before retiring that night to the luxury of his London hotel bed, Rosser confided to his diary: "I accomplished the best piece of work today that I have ever done. I succeeded in getting the army to allow us to buy food and cigarettes from the army depot G-20 at Burton-on-Trent for our men. I am the happiest man in London, I suppose. This means so much to keeping our job going strong. I won't mind working hell out of the boys when I know they are getting plenty to eat."

To celebrate the occasion, Rosser and a new found friend, Sparky Turner of the Embassy staff, went out on the town and to a Broadway production then showing in London. Rosser was too thrilled to remember or care what show he had seen but he did not forget also to note in his diary, "We are all very

5 This was related by Shepard in a telephone conversation with the author in 1969.

thankful and realize more than ever how good a country the USA is."

Under date of July 15 Rosser wrote, "This was a red letter day for Walker and me. We went to Burton-on-Trent, G-20 depot and loaded the car with good USA food and American cigarettes for the boys. They were all as excited over receiving the food as I used to be over Santa Claus' arrival." Rosser again carefully wrote "we are all very thankful."

Don Walker wrote to Colonel Irish a few days later to thank him for efforts expended in having General John C. H. Lee intercede on the ration problem.[6] He emphasized how much the men had improved in spirit and well-being. He assured the colonel that if he would revisit Kelham, he would not be besieged with such tales of woe as he had listened to on his previous visit. Walker reported that the colonel in charge at the supply depot and the major in charge of rations were most cooperative, that arrangements were made to pick up rations on a weekly basis and also to draw supplies from the post exchange such as personal necessities and cigarettes.

[6] "Donald E. Walker to Colonel C. G. Irish," July 22, 1943, *E* file.

XV.

Now, Every Drop Counts

The additional food rations received from the army really worked magic. An atmosphere of good feelings and tranquillity reappeared at Kelham Hall and in the field. The pick-up in the work of the men was immediately noticeable. Walker commented that the quick change in the boys demonstrated the truth of the old adage, "the way to a man's heart is through his stomach." Since the men were on twelve-hour tours, an extra meal was served at midnight when the morning and afternoon tours changed.

With the congenial situation now existing at Kelham Hall and in the field, the summer days seemed to slip by at a fast pace. Despite the continued delays in personal mail from home, the well-fed boys enjoyed their few carefree off hours by reading the news magazines and the *Daily Tulsa World.* As one of the boys wrote, "It is good to get the newspaper and see what Joe Palooka is doing!" Spot news on the lounge radio came at regular intervals during the day. What some of the boys did not get from the radio was usually picked up later by conversation with the others. The men were able to keep well informed on the progress of the war.

Personal letters were doubly treasured above the newspapers received from home, especially those received from Lloyd Noble.[1] He had planned to visit their operations in April and again in July, but both trips were deferred. Responsibilities in Canada and the United States were too demanding and letters he received from Anglo-Iranian authorities were so commendatory on accomplishments. He wrote regretfully in September that he would not come to England. It was the first time in his twenty-two years in the contracting business, he said, that he had not been on the scene of operations where some really vital project was underway.

Noble, always closemouthed about classified projects, described weather conditions in Canada the previous winter when the temperature dropped to sixty degrees below zero and motors at the operations were run continuously because once stopped the oil congealed. He mentioned that 1943 was the hardest year of his life—an unheeded warning of untimely death six and a half years later. Many of his crews were on an eighty-four hour week, but he was spending much more time in travel with the added responsibilities of the corporation and community activities.

The welfare of the University of Oklahoma meant much to Lloyd Noble. Through his efforts undergraduates were furnished summer employment and many graduates permanent employment in the Noble organization. He knew of the interest of Don Walker and others in the university, its football team and his nine-year tenure on the board of regents, where he was serving as the chairman. In one letter he mentioned that Robert S. Kerr, governor of Oklahoma, had made his first appointment to the Board of Regents—Don Emery, counsel for the Phillips Petroleum Company—and added: "I can think of no man in the history of the state that has ever come to the Board of Regents with a better background than Don."

He enlivened his letters with tidbits of state news, such as being in the middle of the fourth bond drive and the appear-

1 Excerpts of letters from the carefully dictated notes of Lloyd Noble to Rosser, Walker, *et al*, mailed from Ardmore or Tulsa, Oklahoma, or Edmonton, Canada, September 30, October 4, 1943, and January 28, 1944, *E* file.

ance in Oklahoma of Red Skelton and other stars from Hollywood. He wrote that a few weeks before, Red had said on his radio program that the difference between a rich Okie and a poor one was that the rich one had two mattresses tied on top of his car. Apparently, the broadcast was made before Red knew he was coming to Oklahoma and this joke left him in the proverbial spot. But Red came out of this pretty well: he told the Oklahoma audience that his scriptwriters had written it in and he just read it off without thinking because he was from Indiana and didn't know what a mattress was! Noble added: "Governor Kerr is supposed to have met him at the airport with two mattresses tied on his car—I wouldn't know about that."

Noble sensed the boys would be interested in his name dropping of fellow employees and acquaintances and, although extremely reticent about the Canol Project, he let them know that George Kinmitz was in Canada along with Herbert Branan, Dewitt McGraw, and Bill Smith; and Louis Priddy, well-known Ardmore restaurateur, was the chief commissary steward. Noble related that L. E. Steward and his wife had been to Ardmore for a visit and had caught a nice string of fish from Lake Murray, that L. E. had suffered a mild heart attack and that Dutch Jensen had gone to Midland to help with West Texas duties. He thought that Ed Holt had been working under too much pressure and he was going to force him to take time off for a few weeks to go to his farm to get away from the terrific work load he had been carrying. In another letter Noble told them his secretary had been hoarding sugar rationing coupons "and one of these days will send enough candy for the bunch. She is figuring on making it herself, so I don't vouch for the results."

Activity in the European theater, meantime, was accelerated with British-American air raids directed at the Ploesti oil fields and refineries in Rumania and night and day bombing runs over strategic war plants located in the Ruhr Valley and industrial cities of Germany. Sea-borne landings on July 10 under continuous fighter protection were made on Sicily in an initial assault which involved nearly 3,000 ships and landing craft, 160,000 men, 14,000 vehicles, 600 tanks, and 1,800 heavy artillery pieces. General Sir Howard Alexander, Eisenhower's deputy in charge of the operation, reported his forces in the joint Anglo-American venture were opposed by 315,000 Italian and 90,000 German soldiers. The campaign ended after thirty-eight days. The enemy lost 167,000 men of whom 37,000 were Germans and the Allies listed 31,158 killed, wounded, or missing in action.[2] After the collapse of Sicily on August 18, the Allies commenced a series of heavy air attacks up the Italian boot to Rome and beyond. Railway terminals and other strategic targets in the Naples and Rome areas were aggressively attacked through July and August. On July 25 Benito Mussolini resigned as head of the Italian government. Marshal Badoglio replaced him and declared the twenty-one-year rule of the Fascist party at an end.

The build-up of military supplies at the south and west ports of Great Britain continued. A principal topic of conversation among British civilians was the guessing game concerning the place and time that the Allies would land invasion troops across the English Channel. All agreed such action would be the next big fireworks of the war.

Rosser continued to take advantage of Shepard's invitation to listen in on the bi-monthly meetings of the petroleum board when he could arrange to be in London on those dates. He felt extremely proud of the work he was directing in the British oil fields and highly honored to listen in on such important and highly confidential meetings. Two matters that had been discussed were of great interest to Rosser. The first was Pluto, the code name for the project involving the laying

2 Churchill, *Closing The Ring*, 23–42.

of a gasoline pipe line from the south of England across the channel to the European mainland. Geoffrey Lloyd had visualized such a line as far back as the dark days of 1941. The matter was now being considered again. The laying of the channel pipe line in the summer of 1943 seemed out of the question. Vessels moving in the channel were under the constant watchful eye of enemy reconnaissance planes and frequently the target of German bombers.

The cross-channel pipe line problem was soon solved by the ingenuity of the oil industry.[3] A two-inch hollow cable, capable of withstanding high pressures, yet sufficiently pliable to be fed off a huge reel from a ship similar to the method of laying the great Atlantic cable many years before was devised. The operation was commenced in the early hours of darkness on a spring night when the moon was down. The next morning, with the job completed and the cable ready for testing, nothing was visible to observers from the air except an unobtrusive abandoned building on the Dover beach that was actually being used as a pumping station. The pipeline, of course, later proved to be invaluable to the success of the invading troops during the coming June of 1944. In fact, twenty such lines were finally laid across the channel.

Another project of great interest to Rosser was the fog fighters project with the code name Operation Fido.[4] Late in 1942 the prime minister had requested the secretary for petroleum to give consideration to some method of dissipating the heavy early morning fogs that closed in on the many air fields now being used by both American and British planes for the day and night bombing of the industrial centers of Germany. After many days of study and discussion, a plan

[3] Longhurst, *Adventure in Oil*, 128–31.
[4] *Ibid.*, 131–34. Churchill in memoranda to Air Marshal Sir A. W. Tedder, October 3, 1943, and to Geoffrey Lloyd, secretary for petroleum, April 1, 1944, commented on the successful use of the fog-dispersing equipment. See Churchill, *Closing The Ring*, appendix, 667, 703.

was designed and engineered to place distillate gas flares along both sides of the runways with a sufficient BTU output to radiate heat to approximately one hundred feet above the runways.

During the spring and summer of 1943, the oil-field workmen had seen American and British planes that had limped home after enemy raids only to crash on their own runways. The use of the distillate flares was an excellent example of the many ways in which oil was serving the military on the fighting fronts. The Americans took great pride in the fact that oil from the British fields was going by pipe line and tank car to refineries at Liverpool and in the south of Scotland.

The small well pumps that British civilians had so aptly nicknamed nodding donkeys and other field equipment, including tank cars and trucks, continued to be protected from air raids of the enemy by fresh coats of green paint that blended with the green of spring and summer pastures of the English countryside.[5] The heavy foliage of the great oaks, beeches, and yews formed a natural camouflage so that the secret operations were not noticeable to a pilot who did not know that a producing oil field actually lay hidden below the big trees of Sherwood Forest.

Despite the plentiful rations now coming to Kelham Hall, all was not serenity for Rosser and Walker. Two of the boys, prone to arguments and occasional fighting, continued to be troublemakers and were finally sent home. H. G. Hobbs, one of the most experienced drillers in the group, became dissatisfied and had given notice that he was going home, much to Rosser's disappointment. Rosser made a strong effort to persuade Hobbs to remain on the job, but to no avail. Bob Webster's broken arm was not healing. On advice of the doctors at American General Hospital in Mansfield, it was decided

[5] R. S. Adams of the D'Arcy safety department had pinned the name "Nodding Donkeys" upon the pump jacks, and natives of the area popularized it.

he should be sent home with Hobbs. Long and Purdy, recruited by the Noble office in Tulsa, arrived in England the middle of July to fill the two-man shortage existing when the crews left New York in March. With these replacements, the record of well completions was being continued, but the lack of substitutes in case of injuries or illness kept Rosser, Walker, and Southwell nervously watching the daily pay roll reports to see if the rigs and trucks were sufficiently manned for maximum efficiency.

By the end of July the American crews had completed thirty-six wells. Rosser calculated that, barring unforeseen accidents and unusually long delays, his drilling crews would easily complete the hoped-for one hundred wells within the twelve-month period, as originally contemplated. Noble, Porter, and Holt appeared to be well pleased with the progress. Rosser continued to receive encouraging and congratulatory letters from the three of them. Rosser wrote Holt that Southwell had told him not to worry about running out of well locations. The D'Arcy Company had a number of additional locations on good-looking geology that Southwell was anxious to test when rigs could be spared from the proven areas. The change from the five-acre well spacing pattern to a two-and-one-half-acre pattern about doubled the available locations. The program would, to some extent, also cut the time required for skidding rigs from completed wells to new drilling locations.

The routine existence for the boys at Kelham Hall was pleasantly relieved from time to time by the comings and goings of visitors. On the weekend following receipt of the first supplemental army rations, they were honored by a visit from Southwell and Jameson with their wives at a Saturday lunch. When Jameson heard that Rosser was planning a trip to London, he was invited to ride in the Jameson car to the suburban area of London in which they lived.

Rosser remembers this occasion with mixed emotions. He

could hardly believe his eyes when Jameson, president of the great Anglo-Iranian Oil Company, knocked on his bedroom door to serve him Sunday morning breakfast in bed. Later, his newly polished shoes were brought in by Jameson with the announcement that the household was now without servants, and he was substituting for the lost help as best he could. Rosser's chagrin at such proceedings rendered him speechless. All he could think of at the moment was that it should have been he who was polishing Jameson's shoes. Rosser was happy to get an early train into London on the pretext that he was due to meet friends at the American Embassy.[6]

It was about this time that Smith D. Turner, Rosser's new friend from the American Embassy staff, paid him a visit. Turner had helped Rosser celebrate the occasion when he had secured the order for the American army rations a few weeks before. Many years later, Rosser received a copy of an article written by Turner commemorating his visit to Kelham Hall:[7]

Can you imagine being shown—with enthusiasm, yet reverence— through the second oldest church in England, by a guide with a western twang in his voice and outfitted in a wide-brimmed Stetson and embossed, high-heel boots? Not in a Hollywood modern-dress version of Twain's "Connecticut Yankee," but actually on the spot, and not as a gag, or in significance of anything special?

As you might guess, it all arose from oil. Until a few weeks ago it was kept very secret—but an item had appeared in a US publication that Oklahoma crews were drilling for oil in the British Isles. This was quoted in the English press—so now I can tell something of my visit with the drilling crews and the unique location of the operations.

The location and production should still not be pinpointed, but it can be said that by US standards it's not much of an oil field.

6 Rosser's diary, September 3, 4, 5, 1943.
7 Smith Turner, *My Guide in Boots and Stetson*, a typewritten copy mailed to Rosser. The article may have appeared in one of the publications of the Anglo-Iranian Company.

Although in the beginning both U.S. and British geologists had held high hopes that a major reserve of oil had been discovered, Britain's need for oil was so great that the production proved to be of immense value to the war effort. With things as they were, it was, of course, worthwhile to develop it to the greatest extent and rapidly as possible.

The dumping down of a colorful rootin', tootin' crew of southwestern roughnecks in a quiet and ancient English countryside has given rise to situations and contrasts fully as picturesque as those imagined by Twain. The English get along with them fine, and not just because they have had to, like it or not. One saving feature is probably the fact that the English have always pictured Americans, particularly those from beyond the Hudson, about as these crews are. The roughnecks were also most carefully selected and impressed with the problem facing them in getting along here. But in addition to this, they have taken a genuine liking to the English, and are deeply impressed with the history and traditions of the neighborhood.

Whether it is so blessed to a greater extent than most areas of England I don't know, but the neighborhood of the oil field abounds in scenes, building and ruins of great historical interest. The high points of all these have been absorbed—more or less correctly—and I was, to my amazement and delight, rushed from one to another with an enthusiasm on the part of my guides, who were but the vicarious owners of these past glories, far beyond that which the natives, whose forebears accomplished it all, could ever have mustered. On a visit to one of the large and very ancient churches, nearby, my Oklahoma guide, with high heels clicking on the stone floor, Stetson swinging in his hand, and voice reverently reduced to stage whispers that echoed from the high vaulted roof, explained with all earnestness the story of the many things of interest he had learned from the attendant a few weeks before, stopping here and there in his discourse to apologize for his inadequacy (". . . we'll have to get the old boy to tell us all about it . . .") ("Boy, there's history in that, but I don't know how it goes . . ."), etc. While their every action and word thus showed interest and appreciation and reverence for the wealth of history and tradition around them, I am sure they would be the last to admit they felt this way, and if so accused, would hoot at it all to show they were not impressed. But this deeper feeling, of course, cannot help but manifest itself, and I feel must account for some of their good relations with the inhabitants of the neighborhood.

But their acceptance by the English rests also on an entirely different factor—they have upheld in every way the glowing stories of the miracles of American achievement by their sheer performance. In cold figures, English crews had been drilling in the area for some months, averaging about eight weeks per well. The Oklahomans proceeded to complete their wells to the same depth in *one* week. Of course, someone here wanted to know "how come?" and the English have now speeded up, and complete theirs in about five weeks. Perhaps even more impressive to the populace than this performance record are their methods of moving heavy equipment. They take down, move, and erect a steel jackknife mast in twelve hours. Most spectacular, I judge from stories heard, is a technique they have of loading and unloading heavy equipment on a very large flat bed truck, with the use of a self-loading winch, the truck is caused to rear its front wheels high into the air, like a great dragon, as the back end is tipped down to allow the tractor to be pulled on and off. The first time this was done was on a crowded dock, and great cries of alarm arose from the onlookers who did not know this was supposed to happen, and thought a terrible crash impended.

All this, on top of the fact that they are Americans, westerners, and roughnecks, has caused them to be a cocky, hell-for-leather outfit, but with exceedingly good morale. They, of course, gripe like hell about everything; but the griping had a verve and a good ring in it. I tried to corner one roughneck, and find out just what the worst gripe was. The best I could get was that the weather was not entirely satisfactory. "Well, I tell you—I like the country, the people are fine, and we are getting plenty to eat now—but who the hell wants to have to sit in front of a fire in August?" From which one must conclude that things can't be too bad.

This morale will stand them in good stead before they are done. They are working hard—twelve hours a day, seven days a week—84 hours per week—with a day off every two to four weeks, and not much to do or many places to go when they do have a day off. Mail from home is slow, and unlike the Army, they are a small group, and there are thus few from home to talk to. And a winter

191

with few hours of light each day is just ahead. So it is good they have things to be cocky about, and that they are cocky about them.

While the facts related are sufficient to make this a gem among the things I have seen, this gem has yet another facet—one even more bizarre, and one that I am sure makes Twain rest uneasily in his grave, in self-reproach at lack of imagination in creating incongruous situations.

These Texas and Oklahoma roughnecks are staying in a monastery with a group of Anglican monks! complete with robes, sandals and skull caps!

It is true the crews have a wing of the monastery to themselves, and have their own separate mess, but they are in daily contact with the monks. The question has often been asked: "Will the monks succeed in getting the roughnecks to take the vows, or will the roughnecks bring the monks out and help with the job of setting casing on the oil wells?" Actually, though they see each other often, the two groups keep pretty well apart, and have maintained the best of relations. The Brothers have given them a little ground and in their spare time the roughnecks have made a little garden, and also cut a supply of firewood.

The roughnecks are most enthusiastic about the Monks' establishment (as they were about the church) and arranged to have me taken into the chapel (which is, surprisingly, an amazing example of ultra-modern architecture) by "Brother Edgar," a delightful old fellow—almost blind, with a long beard, who, to quote the driller, has something to tell about every brick and stone in the building. We gladly left some coins in the poor box for the building fund and the charities supported by the monks.

And so we have the picture of a monastery with high-heeled boots standing in each room, with huge breakfasts of oil-field food, with talk of lost circulation, new locations, cementing, and of what O.U. and the Aggies will do this fall on the football field; and with it occasional all-night crap games after payday, with pound notes piled in heaps like fallen leaves. Beside these things, such points as drilling in a blackout, where you find the mud pumps by falling over them, and the mud pit by stepping into it, and the location being so situated that hundreds of bombers go over low almost every night, drowning the noises of the rig, are mere oddities which no longer excite the crews.

At the end of June, Rosser and Walker reviewed their work since arriving at Kelham Hall and on July 8, Rosser wrote Holt giving him a good picture of the situation as it stood at the time. He mentioned a recent visit to their operations by Sir William Fraser, chairman of the Anglo-Iranian Oil Company, B. F. Jackson, the company's New York representative, and Sir Edward Packe, government representative.[8] The distinguished visitors witnessed drilling operations, shook hands with the men and complimented them on their work.

Rosser then discussed some of the technical problems encountered, no more hazardous, he said, than in Illinois—"Just set her on bottom and turn her to the right." He mentioned that a lost circulation problem in the surface formation was resolved by digging a central mud storage pit, that "when we lose mud we just start pumping from the central reserve and keep on drilling."

He described the formations as clay and red beds or red marls for the first four hundred or five hundred feet, then five hundred to six hundred feet of red sandstone followed by a two-hundred foot section of limestone, then shales and a little limestone "to the pay" that they were using three new bits and two retipped ones on every well.

Rosser and Southwell toured the territory on July 10 on which the D'Arcy Company had done considerable geophysical work and looked at the locations that were approved for outside exploratory wells. Thirteen such locations were drilled before the contract period terminated. On the agreed twenty-five hundred feet of drilling to be treated as one well, Noble and Fain-Porter companies were credited with twenty-three completions for the thirteen outside test wells.

The outside drilling proved very disappointing. All were dry and abandoned except the small production found at Nocton. The Nocton number 1 well was finally plugged and

8 "Rosser to Holt," July 8, 1943, E file.

abandoned because of a bad fishing job. Twelve joints of drill pipe were left in the hole, but the other three Nocton locations were completed as small producers. But now that the across-channel invasion date was growing closer, every drop counted. The oil being produced in Britain at this time was beyond money value.

Material and replacements of equipment lost at sea were still arriving at British ports on the west coast. Rosser and the truck drivers were busy picking up these arrivals. Rosser's diary entries from July 20 to 23, inclusive, give a good picture of the situation.

July 20, Rosser wrote: "Arrived in Liverpool at 11:30. Checked in at the Adelphi—room 346. Went to Arbuckle Smith and to docks. Cleared the arrival of Mr. Long and Mr. Purdy's baggage. I just gave up the two boxes of cigars and the cigarettes they had in their trunk. Three cartons cigarettes and one hundred cigars, they wanted $30 worth of duty. I could not see the point."

July 21: "I got up at 4:30 A.M. this morning and caught the 5:05 train for Nottingham. I changed trains three times and finally got to Nottingham at 10 A.M. I drove like hell to Bottesford to meet Mr. Southwell and a Mr. Jackson (refinery) and Mr. Confield with the Burma Oil Co."

July 22 relates: "I took the K-8 and Sikes K-7 trucks and left for Cardiff, Wales at 5 o'clock today. We arrived in Cardiff at 2 A.M. Just in time for the sirens to go for an air raid alarm. We checked in the Queens Hotel and everybody was getting up and going to the air raid shelter. But like all crazy Americans, we went on upstairs and went to bed. The guns went off and woke us up and finally the all-clear sounded at 4 A.M. No bombs."

July 23: "We got up at 8 A.M. after hardly no sleep and went to the docks to load the drilling mast. It was covered up with USA box cars. I had a hell of a time getting all of it loaded, but finally at 5 P.M. we pulled off the docks and went back to the Queens Hotel for a good night's sleep but left orders to be awakened at 5 A.M."

The arrival of the replacements and equipment received was not sufficient to rig up the fourth National 50 rig. Despite this disappointment, Rosser was able to report to his home office in mid-September that sixty-four producing and twelve dry wells had been completed, that they were still operating one outfit with A. C. electric motors and some junked drill pipe, "averaging two twist-offs per well with this pipe." He was very complimentary of the bang-up job Gordon Sams was accomplishing as tool pusher.

Military bases and supply depots, military personnel, supplies and equipment were beginning to crowd Northern Ireland and Scotland, Wales and England, and troops from the United States were a common sight in the cities and towns of Britain by mid-summer 1943. Rosser, Walker, and the boys had made many friends among the young men who seemed to be everywhere. The first American soldier spotted by some of the boys in Nottingham was a black man from Alabama. To celebrate the occasion, they had brought him to Kelham Hall for dinner. Rosser commented: "That American soldier dressed up with a Stars and Stripes patch on the shoulder sure did look good to all of us."

Rosser and Captain Hull from Olney, Illinois, who was billeted close by to the south of Eakring and the Duke's Wood area, met and became close friends. On their first meeting Rosser invited the captain to Kelham Hall for a drink of Bourbon. On seeing the Bourbon, Hull related how small the amount the officers club was getting at his base, although they had a large surplus of Scotch. He suggested he would be happy to trade two bottles of Scotch for one of Bourbon. He made a show of a considerable supply of British pounds he continually carried, he said, for the purpose of purchasing any liquor he could find. Rosser recalled that American Embassy personnel in London were always in short supply of Scotch.

As always, Rosser was immediately alert to the strategic position he and Walker were in with respect to liquor. So without delay, a deal was struck between Rosser and Captain Hull whereby the boys' whiskey stock received two bottles of Scotch from Hull for each bottle of Bourbon. Half of the Scotch thus acquired was taken by Rosser on his trips to London and there traded to the American Embassy boys at the ratio of two bottles of Bourbon for one of Scotch delivered. In this manner the Rosser-Walker combine was fast building up a good supply of both Bourbon and Scotch. The liquor trades continued until the work of the American drilling crews was completed in England to the satisfaction of all concerned. At the end Walker and Rosser finally gave the accumulated liquor surplus to their British friends at the Eakring and Burgage Manor offices and in the Kelham village area.

Replacements for assembling the fourth rig were now coming through, it seemed, at a faster pace. The arrival of cargoes was stepped up, due to the success of the British and American navies against the enemy's submarine attacks in the Atlantic. The surprising accuracy of their depth charges on U-boats the navies laid to their use of sonar listening devices.

Another incident at this time that brought excitement to Kelham Hall was the news of the arrival at Liverpool of a box containing forty pairs of badly needed safety work shoes. The shoe repairman in Newark had about exhausted the repairs that could be done on the originals. One of the boys, still active as an employee of the Noble Company in the Gulf Coast division, remembers it took several days of fittings and tradings to get the right size for each man. He is doubtful today that each man got the shoes intended for him. Another of the boys says he remembers Phil Albritton claiming he was doled out two shoes, both of which were built for the left foot! A few hours on the shoe stretcher at the shoe shop in Newark, he thinks, remedied the misfit.

At one period in the early summer of 1943, Rosser wrote Holt that two of the outside wells then being drilled were approximately fifty miles apart. The time-consuming travel of the crews to and from these locations was defeating the completion schedule that had been so greatly admired and applauded by the companies, but the travel time meant little to the boys because now finding more British oil was their most serious business. This situation was soon remedied when it was decided to move the rig on the outlying location to a drill site near the village of Eakring in the center of the producing area for a deep test well.

The day-to-day events relating to the boys' work through August, September, and October were the usual and ordinary ups and downs that any oil-well drilling crews might expect under similar conditions in the United States. The humdrum of twelve-hour tours, seven days a week, was interrupted by a number of trivial, but interesting matters.

For example, Rosser reports having been arrested by a woman policeman for speeding through the town of North-hampton, while on a hurried-up trip to get bearings to repair a burned-out crank shaft. On August 11 Rosser's diary carries the philosophical entry "there is no telling what will happen next, after a man gets arrested by a woman. She was very nice, but anyway, I will hear from the court." A few days later, Rosser got notice he had been fined one pound and mailed a one-pound note to the magistrate's court.

On August 10, two trucks and drivers left for Glasgow to pick up equipment that was arriving there. During a two-day delay in discharging the ship's cargo that was pouring into British ports, Rosser reports he got a quick look at Loch Lomond and surrounding scenic areas.

The latter part of August was busy. Getting passage to America for Boucher, Purdy, and Webster, who was still having trouble with his arm, was difficult. Near the end of August, Thomas Cook & Sons with the help of the American Embassy and the assistance of Mr. Southwell in getting releases from the Ministry of Labor, passage was finally secured. On September 9 Rosser went with Bob Webster to board his ship; the last of the four men to be returned to America.

Routine completions, lost circulation, twist-offs of well-worn drill pipe, and other minor troubles were now consuming the drilling crew's time. Occasional trips to west coast ports and to London broke the monotony of the daily routine. Mail from home continued to be slow. The arrival of news magazines and the *Tulsa World* was good but poor substitutes for the hoped-for letters from home. The return of Hobbs, Webster, Boucher, Edens, and Purdy to the United States left a noticeable vacancy at Kelham Hall. Pete Oaks spent a few days in the hospital because of food poisoning. Virgil Latham also was hospitalized several days because of trouble with his stomach ulcers. Glenny Gates had dived in the mud pit when German bombers flew low over the rig location and had to be brought into Kelham Hall to be cleaned up. It was fun for everyone except Glenny. He suffered no ill effects because the rig was operating with a plain water and dirt mixture that at the time contained no chemical additives. During this period, Rosser and the drillers were busy teaching a number of D'Arcy men their drilling procedures on the National 50 rigs.

Late in October, a Major Armstrong, former drilling superintendent for Burma Oil Company now serving in the Indian army, paid the Kelham area a visit. Armstrong had been one of the men who helped destroy refinery facilities and plug producing wells in Burma before Japanese occupation. He was keenly interested in the American equipment and drilling procedures and spent two days observing operations of the American crews.

Rosser and Walker reported a bombing raid over the Nottingham area on October 31. Perhaps the most interesting event at this time was the decision of Rosser to discharge the British cooks and stewards after a conference with Mr. Southwell and make arrangement for replacements. The matter was brought to a head by another fight between one of the boys and the kitchen help. Captain Hayward, the owner-operator of the Fox Inn, had previously given Rosser some information about the drinking habits of the cooks and stewards in the afternoons when they should have been on the job at Kelham Hall.

The exact date of the incident does not appear to have been recorded, but subsequent events of that day will linger in the memory of Rosser, Walker, and the boys. In the afternoon Rosser and Walker had made their weekly trip to Army Depot G-20 at Burton-on-Trent. During the return trip, Rosser had casually remarked to Walker that they would have to get back to Kelham early, because he and Walker must do the cooking for the evening meal. Rosser explained that at noon he had fired the entire kitchen crew and had given them until five o'clock to pack their bags and be out of their quarters at Kelham Hall.

Walker's reaction to the news that he and Rosser were going to do the cooking was somewhat explosive. As Rosser tells it today, Walker with appropriate gestures told Rosser he had not come to England to be "Mother Superior" to a bunch of roughnecks and since Rosser had run off the only help they had, he could just damn well do the cooking until they got replacements.

It appears that Rosser and Walker, however, with help from some of the others managed the evening meal. After dinner Rosser asked for attention, the usual procedure when announcements were made affecting the group. He explained the kitchen help had been discharged and a new crew would be on duty the next day. In the meantime, he wanted them to know that Walker had been promoted to the status of "Mother Superior" of the monastery, and the boys were to govern their conduct accordingly.

The event caused great hilarity at the moment, but no one envisioned it would become the permanent designation for Donald Edward Walker by friends through his many years of service to the Noble Drilling Corporation. To this day, the great friendships and admiration of the boys now living, their children, grandchildren, and the long list of Noble employees who have come and gone continue affectionately to address Walker as "Mother Superior Walker."

The American roughnecks were all the while endearing themselves to the kindly folk of the English Midlands. The long days of the pleasant summer passed swiftly. September and October came and left almost without notice. There was a feel of the coming autumn and winter in the air. November, with the American Thanksgiving date ringed in red, showed on the big calendar Walker had recruited for the lounge. The calendar served as a memorandum for coming events and the dates that some of the boys were now having with a few of the friendly English maidens of the neighborhood. Saturday nights at the Fox with guitars and harmonicas the boys had dug out of their luggage were always a highlight for the morning tour crews.

It was a crisp evening early in November that Robbie Robinson had brought his truck to a stop at Kelham Hall to permit Rosser to alight and head to the dining room for a dinner of some of that good U.S. chow. On entering the lounge, he found Walker busily engaged in serving a drink to a military visitor wearing the insignia of a major. He introduced himself as Major R. C. Moore, and said he was attached to the SOS (Services of Supply) unit at the nearby Mansfield General Hospital. The young major was on his way to meet friends in

Nottingham and seemed to be extremely anxious to get along. But after a refill of his drink and one for Walker and Rosser, he explained that Major General Lee's office in London had requested him to contact Mr. Rosser and arrange for the inclusion of a visit to the oil fields for General Lee during his inspection trip to the hospital at Mansfield in the near future. The general, he said, was quite anxious to see the drilling operations being conducted by the American drilling crews in the Eakring and Duke's Wood area.

Both Rosser and Walker expressed delight at the possibility of a visit to the field and perhaps to Kelham Hall from General Lee. They assured Major R. C. Moore they would cooperate in every possible way to show the general as much of the field operations as he might desire. Rosser had urged General Lee to pay a visit to Kelham Hall and the drilling operation when he had obtained the order for U.S. army rations. He had not really thought that General Lee, one of General Eisenhower's top aides at Supreme Allied Headquarters, would actually take time to make the trip to Kelham and the oil fields. Major Moore said he would stay in touch with Rosser and Walker and let them know the details concerning the general's visit, when he had received further information.

XVI.

Sunshine and Shadows

"Today has probably been the most important day of my life," Rosser wrote in his carefully kept diary late on the night of November 11, 1943. He went to bed with his memories of the events of this important day and visions of a possible military career under the command of Major General John C. H. Lee, Commanding Officer of SOS (Services of Supply) European theater of World War II.

Little did Rosser know that within forty-eight hours, he would write in his diary: "This has been the most tragic day we have had since our arrival in England."

On the evening of November 10 Major R. C. Moore called Rosser to confirm detailed arrangements regarding the visit of General Lee. By prearrangement Rosser arrived promptly at 11 A.M. in the village of Ollerton some twenty miles to the south and east of Eakring. As the well-worn Plymouth braked to a stop near the village square, Rosser realized that an Armistice Day memorial service was already underway. Leaving the car in charge of one of the truck drivers who was familiar with the roads of the area, Rosser walked into the square to see the maneuvering of the color guard and to hear the remarks of Captain Emanuel Carlsen, the chaplain attached to the American General Hospital at Mansfield. Carlsen was closing the ceremony with prayer as Rosser saw Moore approaching. General Lee was asking that Rosser be brought to his car. Lee's staff and aides occupied eleven open army cars, while Lee was seated in a black open Cadillac. The cars formed a column around the square behind the general's car which bore the insignia and the flags of a major general. Approaching the car, Rosser was invited by Lee to get in and sit in the back seat with him.[1]

The sun had failed to disperse a damp fog that day. A stiff breeze with a threat of winter in it was coming out of the north. General Lee, noting Rosser's seemingly inadequate field jacket, inquired if the weather was too cold for him? Rosser, anxious to be agreeable and not cause any delay, promptly answered "if it is not too cold for you, General, it is not too cold for me." The general insisted, however, on sharing with Rosser the heavy lap robe that covered his field boots. Thus settled, Lee pulled his overseas cap down against the north breeze and told the chauffeur that "we'll make the Mansfield hospital first."

Two other smaller hospitals were also to be visited, but the general said that these stops would be delayed until after the inspection of the Mansfield hospital, where they would have lunch. During the trip from Ollerton to Mansfield, Rosser took the opportunity to express the appreciation of his drilling crews for the use of the hospital as well as the canteen privileges they had been accorded by the commanding officer and staff members at the hospital.

The inspection of the hospital seemed perfunctory to Rosser except for one incident which he never forgot. General Lee made a point of raising the lid of several large garbage cans. At one he hesitated, reached into the garbage, and withdrew a long loaf of bread, still wrapped in its cellophane jacket, and, to the amazement of Rosser and others trooping along in the wake of the fast moving leader, ripped open the bread con-

[1] Rosser's diary, November 11–12, 1943, and taped interviews.

tainer and broke off a large chunk from which he took a
healthy bite. Turning to the red-faced officers of the hospital
staff, General Lee had a few words to say about unnecessary
waste in the army kitchens. Rosser thought how good it would
have been for Littlejohn to have heard Lee's remarks when
he was in London begging the army for food rations to feed
the American roughnecks.

After a hearty but simple lunch in the officers' mess in the
hospital, Rosser was relieved when the cars moved back to
familiar territory at Eakring. During lunch Rosser had observed more military brass than he had ever seen at one time
and had noted that the lowest in rank were two young captains.
He was pleased that Littlejohn was not on the inspection trip
because he had no desire to renew his acquaintance with him.

Hard rains had fallen the night before, and Rosser knew the
field roads would be rough and slick. At his suggestion, the
driver transferred to an army jeep for the trip through the
field. Rosser and the general climbed into the back seat. Rosser
directed the driver through the winding muddy roads of the
Eakring and Duke's Wood fields. He felt these areas would
give a good example of the drilling operations. At the top of
the ridge that divided the Duke's Wood field where the mud
did not appear to be deep, Rosser asked the driver to stop and
inquired if the general would like to go on the rig floor of one
of the wells. He answered by springing from the jeep to a solid
spot on the side of the road. Rosser followed and led the way
to the nearest rig equipped with a jackknife mast, powered
by two 150-h.p. electric motors, and presenting a typical
operation.

The roughnecks, although visibly surprised to see a major
general coming on the rig, continued their routine of work
like well-disciplined soldiers. Rosser explained the advantages
of this type of equipment over the old conventional drilling
rigs and gave a quick demonstration of lowering and raising

the drilling mast. He explained how the rig was skidded over the surface on the wooden mat built for that purpose to the next location which saved many hours of delay in tearing down and rebuilding the derrick as required with the older conventional types of rigs. The unitized draw-works also, he pointed out, saved much time in rigging up the equipment. General Lee seemed extremely interested and moved about the rig floor shaking hands with the boys and asking a few questions on his own.

At last Lee indicated the party must be moving along and invited Rosser to continue the trip with him. He accepted the invitation without hesitation, but urged the general and his party to take a few minutes to visit the nearby Anglican monastery at Kelham Hall where the workers were billeted. Perhaps the fact that General Lee was an active lay member of the Anglican church had something to do with his telling the driver that they would stop at the Kelham monastery. He requested Rosser to direct the driver by the best route. Within minutes the general's party was being parked along the driveway to Kelham Hall where they were met by Walker and introduced to Father Gregory. Special services had been planned for the visit. It was clear that the general was pleased and impressed by the courtesy paid his group. Following the brief services, which memorialized the Allied forces killed in World War I, Lee led the way out of the small chapel through the hall to the area of the parked cars, but not before he had noticed the chapel's poor box positioned in a conspicuous place near the door. As the general passed, he carefully and deliberately fished a $20 bill from his billfold and slipped it into the box. The officers trooping behind their leader immediately followed suit. Father Gregory, in later years, told us with a wide grin on his face, that never before or since the general's visit, had the poor box held so much for charities in good hard cash money.

As the big car with Rosser and the general now comfortably sharing the heavy lap robe headed out from Kelham Hall to the main highway, Rosser continued his explanation of the drilling operations in the area, the disposal of the oil being produced, and some of the difficulties that the boys had overcome now that they were getting sufficient food. All the while, he was wondering where the general's trip would terminate and how he would get back to Eakring or Kelham Hall. Maybe he had been too hasty in leaving his car and driver at Eakring.

Rosser soon got an answer to his question. The general's driver led the way over the main highway from Eakring headed for the historic cathedral city of Lincoln. The cars entered the town and headed for the railway station and began to find parking places near where a half dozen railway passenger cars waited on the siding back of the station. The general boarded one of the cars and was immediately followed by the others, including Rosser. The car which Lee entered was fitted up as a lounge, complete with a bar at one end. Rosser had noticed that the rail cars were equipped with telephone and telegraph connections. As the various officers arrived in the lounge, they gave the general reports of various military activities and happenings of the day, reading from notes made in small looseleaf notebooks, one of which each officer carried. Rosser says he sat "bug-eyed" as the officers made their reports. At their conclusion, the general told one of the captains standing nearby that he would have Scotch and water and asked Rosser what he would like to drink. Rosser says that for the first time he found out why those low-ranking captains were part of the major general's group. It seemed the captains were brought along to staff the bar with a couple of good bartenders.

Drinks were served all around and conversation about the happenings of the day was buzzing loudly. The general moved nearer Rosser and without preliminary conversation suddenly asked if he would be interested in joining his—the general's—

outfit, if and when Rosser would be relieved of his duties in the Eakring oil fields. The general explained that the Army Service Forces planned to develop an independent water supply for the troops that would soon, he hoped, be moving across Europe. Present planning called for drilling wells to comparatively deep water sands which had already been mapped by people familiar with the geology of western and central Europe. Rosser, somewhat surprised at the sudden question, realized then that the general's visit to the oil fields had been much more than a courtesy call. Rosser assured the general that nothing would please him more than to be attached to any outfit under his command.

Lee received the compliment with the remark that there were good and bad outfits, and a soldier's life in the army many times depended on what outfit he was in. The general thought that several hundred deep water wells might be necessary to supply the many military units that would soon be active on the European mainland. Rosser, still retaining his enthusiasm and the eagerness of youth, told the general that he had previously talked with the officer in charge of the drilling outfits in a storage yard near London. The officer had come to Eakring to discuss with him some of the problems that had been encountered in drilling water wells at military bases in the Midlands and in the south of England. Rosser had discovered that the operators of the army equipment had drilled an $8\frac{5}{8}$-inch hole and were attempting to run $8\frac{5}{8}$-inch casing into the hole, which could not be done successfully. With the bravado of youth, Rosser told General Lee that it was clear the army boys in charge of the drilling operations he had seen were inexperienced. If he volunteered for services in the army, after the completion of his work for the D'Arcy Company, he would want a rank superior to the men now operating the drilling equipment, because he felt this would be necessary for him to direct successfully the drilling operations.

When Lee asked Rosser how old he was and Rosser told him thirty-one, the General inquired of one of his staff who appeared to be well informed on army regulations, what was the highest spot commission that could be given to an enlisted man thirty-one years of age? The officer advised the general that a captain's commission was the highest rating that could be given at this age. The general pursued the question by asking if there was anything in the regulations prohibiting a higher rank being conferred after a man had been brought into the army as a captain? The adviser assured him that after a man was once in the army with a fixed rank, there was nothing to prevent the commanding officer from raising his rank as he saw fit after a period of twenty-four hours.

General Lee suggested that Rosser come to see him at his office when he was next in London. Later Rosser stopped at his office several times but found the General engaged in other matters and was never able to make connections with him before he left England. After the war Rosser learned that the water-well drilling project had been cancelled when it was decided that sufficient water was available to the army after it was chemically purified in treating plants set up for that purpose. The drilling equipment held by the army in England was shipped years later to Houston and offered for sale to the public.

The sounding of the dinner gong was well understood by everyone. The long afternoon trip to the hospitals and to Kelham Hall in the teeth of the wind coming off the North Sea, climaxed by a couple of stiff cocktails, served to build up huge appetites. It certainly had done the job for Rosser. Young Rosser thought southern fried chicken with hot biscuits and cream gravy was a fitting climax to the unbelievable afternoon he had had with the commander in charge of all services of supplies for the millions of Allied forces in England.

General Eisenhower had long recognized General Lee's

capabilities.[2] In later years, he characterized Lee as a man of "stern insistence upon the outward forms of discipline which he himself meticulously observed. He was determined, correct, and devoted to duty; he had long been known as an effective administrator and as a man of the highest character and religious fervor . . . . Indeed, I felt it possible that his unyielding methods might be vital to success in an activity where an iron hand is always mandatory."

The American drilling crews in the field, Rosser felt, would easily complete the one-hundred or more hoped-for wells within the 365 drilling days unless outside wildcat drilling should discover new and larger producing fields. Several wildcat wells had been drilled to depths sufficient to test the known sedimentary formations considered to have possibilities of oil production. One well at Formsby and one at Nocton had been drilled into igneous rock which the geologists felt precluded commercial production in those areas. The deepest of the outside test wells was Nocton number 2 that reached the depth of 7,480 feet. Two wells at Nocton later proved to be small producers. On the whole, the outside wells had been disappointing.

On Saturday morning, November 13, Walker set off in the Plymouth for the supply depot G-20 at Burton-on-Trent to pick up the weekly American army food rations. Rosser with Robbie Robinson driving the K-7 truck headed for the Liverpool docks to pick up a C-100 pump and supplies (another box of work shoes) received by the shipping agent. The day was heavily overcast with thick fog that hung close to the ground. But nothing dampened Rosser's spirits since the visit of General Lee and his invitation to join up with his organization for the big show on the continent which everyone felt would soon be the next big offensive of the Allies. He was explaining this to Robbie as the truck pulled up at the guardhouse gate to the

2 Eisenhower, *Crusade in Europe*, 235–36.

docks where he was to meet the shipping agent. As he climbed down from the truck, Rosser was told that he had an emergency call from Kelham Hall. Walker was on the telephone saying that he had some bad news. Herman Douthit had fallen from the double board of the drilling mast at location 148 in Duke's Wood and had been killed. Robbie was standing nearby when Rosser's knees seemed to buckle a little and the telephone receiver dropped to the floor. The man in the guardhouse pushed a chair under Rosser and picked up the telephone receiver. By this time, Gordon Sams was on the telephone demanding to know if he was there. "Did you hear what Walker said about Herman Douthit?" Robbie told Sams he was on the phone and Sams repeated the information. Rosser and Robbie immediately loaded the box of shoes and the C-100 pump and started for Kelham Hall. They pulled into the courtyard about 9:00 P.M. to find the heavy shades of death had descended on the boys at Kelham Hall.

Later that night, Rosser wrote of the accident and noted that he had rushed back to Kelham after receiving the call on the Liverpool docks.[3] He noted: "but there is nothing I can do—Don and Gordon, as usual, have everything done that can be done."

Accompanied by Southwell the next day, Rosser went to the American General Hospital at Mansfield to see Chaplain Carlsen to discuss funeral arrangements. Rosser then caught a train to London to advise the American Embassy of Herman's death and to discuss the possibility of sending the body home. He got off a cablegram to Ed Holt telling of the tragedy and requested Holt to notify Herman's family.

Officials in the Embassy explained that shipment of the body home at this time was impossible due to wartime regulations. The alternatives were burial or cremation. After talking with

[3] Rosser's diary, November 14–18, 1943; cablegram from Walker to Holt, November 14, telegram to Mrs. Louise Douthit from Ed Holt, November 14, cable from Southwell to Noble, November 16, all bear upon the tragedy, *E* file.

Walker, they agreed that cremation was not acceptable. Again, Rosser went to see his old friend, Colonel Irish. The next day, Monday, November 15, Rosser cabled Noble's Tulsa office that Colonel Irish had arranged the burial with full military honors in the Brookwood Military Cemetery in Surrey, a short distance out of London. Funeral services would be held at 10:00 A.M. at the Kelham parish church. Rosser cabled that if after the war Mrs. Douthit desired, Herman's body would be shipped home. He also expressed sympathy to her from himself, Walker, and all the boys.

At 5:50 P.M. that evening, with a cold fog settled over the city, Rosser left London by train for Newark. Rosser sat alone in the train with a big spray of flowers that seemed to wilt with the weight of the outside rain that trickled down the windowpane of his compartment. The fog and rain of the dark day were most depressing. He arrived in Newark that evening and took a taxi to Kelham Hall to find that he, Walker, Sams, Albritton, the driller on Herman's crew, and Peter Ashcroft, an employee of the D'Arcy Company who was working as an apprentice, had been subpoenaed to appear at the coroner's inquest to be held the next afternoon at 3:00 P.M. in the visitor's waiting room at the Mansfield general hospital. All rigs in the field, including the D'Arcy rigs, were shut down from 6:00 A.M. to 6:00 P.M. in respect to Herman Douthit.

The funeral services at the fourteenth-century Anglican church were, as Walker wrote Mrs. Douthit, attended by Herman's American companions as well as the British friends he had made since arriving at Kelham.[4] In fact, many who came could not get inside the church. The services were conducted by the chaplain from the hospital at Mansfield and the minister of the parish church. The services followed the traditional Anglican form that Mrs. Douthit would find in the American Episcopal prayer book. After a brief address from Chaplain

4 "Donald Walker to Mrs. Louise Douthit," November 26, 1943, *E* file.

Carlsen, using as his text "Father into thy hands I commend my spirit," the funeral cortege proceeded from Kelham to Brookwood Cemetery where the American flag was removed from the casket. Later, Walker had the flag mailed to Mrs. Douthit. The traditional military salute was fired over the grave. The pallbearers, close friends of Herman, were: Joe Barker, Phillip Albritton, Dewey Aycock, Loran Robinson, Joseph Waits, Ray Hileman, Gene Rosser, and Alan Rutherford of the D'Arcy Company. They accompanied the body to the United States military cemetery. Southwell later wrote that, as the service began, the sunshine broke through the clouds over the little church, as if touched by some great magic hand.

At the coroner's inquest, the witnesses gave a clear statement of the accident:

PHILLIP ALBRITTON sworn saith:—

I reside at Kelham Hostel, Kelham, near Newark. The deceased was Herman Douthit and he resided at Kelham Hostel. He was an Oil-Field Worker, aged 29 years. A married man—wife in America. At 11:30 on Saturday the 13th November 1943, I saw deceased go up the derrick at 148 location to do some work. At about 11:45 A.M. I saw him at the foot of the derrick. He was unconscious and was removed to 30th General Hospital, Sutton in Ashfield. Deceased's job was to go up the derrick to tie the line on the platform. Then he would climb higher up the derrick and give orders for the platform to be pulled up and then he would bolt it and then come down the derrick. The platform is for holding the boring tubes. When the platform has been pulled up, the rail fouls the ladder. The weather was bad. Rain in the morning.

(Signed) PHILLIP ALBRITTON

PETER ASHCROFT sworn saith:—

I live at 14 Newlands Estate, Forest Town, Derrick Man. On Saturday the 13th November, I was working on the derrick at

Dukes Wood. There were 5 of us including three American citizens. I saw deceased go up the Derrick about 11:30 A.M. and he went up for the purpose of attaching a rope to the platform. Having done this, he climbed higher and called someone to pull it up. Then when he was coming down to take the catline off, he fell. The platform rail juts out over the ladder. He slipped off the platform rail, not off the ladder. He fell outside the derrick onto the ground. He would be about 55 feet up when he fell. He was picked up and taken to hospital. I had known him only three days. The operation was a normal job. There was nothing unusual about this derrick.

(Signed) PETER ASHCROFT.

RADFORD SMITH ADAMS sworn saith:—

I live at D'Arcy Exploration Company, Eakring, I am a First Aid Attendant. At about 11:55 A.M. on Saturday the 13th November 1943, I received a telephone call from Rig No. 148 Dukes Wood, Eakring, stating that there had been a serious accident, a man having fallen down the Rig. I at once proceeded with the Works ambulance and first-aid kit to the scene of the accident. On arrival I observed the man lying on the ground having sustained serious head injuries (frontal region). He was unconscious. I considered there was nothing to be done other than apply a temporary dressing and proceed with all speed to the hospital. I accompanied the patient in the works ambulance to the United States Emergency Hospital, Sutton in Ashfield. On arrival, I was conducted to the casualty receiving station where the patient was received by Capt. Edwin Clausen who examined the patient immediately and pronounced life extinct.

I arrived at the hospital at approximately 12:30 P.M. and I was accompanied by Phillip Albritton, an American, one of his workmates, who is also foreman driller.

(Signed) R. S. ADAMS

In later years Wallace Soles told of the poignant scene in his office where Gordon Sams had rushed to telephone the hospital that the ambulance with Herman was on the road. Soles related the anguish of Sams as he knelt on the floor, alternately imploring the doctor he had gotten on the telephone at the hospital and God not to let Herman die.

Southwell, Sir William Fraser, and Jameson wrote letters of sympathy and condolence to the Noble and Fain-Porter companies, copies of which were sent to Mrs. Douthit. Rosser and Walker also wrote personal letters to Mrs. Douthit and told her they were pleased with the thoughtful sympathy and friendship indicated by the many floral offerings sent by English friends from Eakring and Burgage Manor. Friends in the village, Rosser noted, were contributing to a small fund for Herman's widow and it sure did make him feel good because "I know how much a few shillings mean to them, but it makes us feel good to know that we have made such good friendships with the English people."

After the war, in accord with a policy of the Veterans Administration, Douthit's body, with the remains of other U.S. service men previously interred in cemeteries throughout Britain, was transferred to the American Military Cemetery and Memorial located near the beautiful university town of Cambridge on the river Cam. The only American civilian interred in the cemetery is Herman Douthit, who is buried in Plot C, Row S, Grave 2. A white Italian marble headstone, like the ones over the American war heroes buried here, marks Herman's grave.[5]

[5] Visited by the author in 1969.

XVII.

Christmas at Kelham

Despite the shadow of tragedy that lingered over Kelham Hall after Douthit's fatal accident, the enthusiasm of the youthful American drilling crews was bolstered by the approaching Christmas of 1943. Walker was busily arranging the details of the holiday season. Since October, he had been reminding everyone Christmas was just around the corner.[1] Kelham Hall, he announced, was going to have a real old-fashioned, down-to-earth, American Christmas, with a Christmas tree, turkey dinner, and all the trimmings, including plum pudding with flaming brandy. As an additional social event, Walker suggested that Kelham Hall should have an open house with eggnog aplenty to repay the many courtesies the English folk had shown them since their arrival at Kelham.

To the homesick boys in a foreign land three thousand miles from home, living under severe wartime restrictions with an occasional air raid warning thrown in, the promise of the festive Christmas was a heart-warming elixir.

During the past several weeks, the boys had been driving pretty hard to make up the time lost because of fishing jobs and other types of well problems. Despite the trouble on the

[1] Frank Porter advised Don Walker, in a postscript to his letter October 28, 1943, to go all out on a Christmas party, "extend yourself as far as you want as to food, drinks, and what-not at our expense," *E* file.

rigs and the delay in receiving some of the much-needed equipment and repair parts, the good record of well completions had somehow been maintained at a pace of which Rosser and Walker felt they could be justifiably proud. The most satisfying thing, of course, were the letters complimenting their work that Southwell and others of the Anglo-Iranian Company had written to Noble and Porter back home.

Another aspect of their operations that Rosser and Walker felt merited commendation was the nearly perfect conduct of their workers. The boys had set up a policing committee selected from the group soon after moving into the Kelham monastery. It had been necessary for Rosser, Walker, and Sams to iron out a few clashes of personalities and the minor gripes that naturally arose under the circumstances and restrictions under which the group lived and worked.

Of course, there had been the trouble at the Saracen's Head in Southwell when one of the boys, who had a few beers too many, seemed to think the bar was the men's room. There was also the time, soon after their arrival, when a telephone at the Clinton Arms Hotel in Newark had been ripped off the wall because the telephone operator failed to get His Majesty King George VI on the telephone so that the man from "Stroud-America" could register some personal complaint. On one occasion some of the boys were taken into custody by the local constable for the unauthorized use of the D'Arcy Company car that was marked exclusively for war work. But in all such minor incidents the authorities had been most understanding. Payment for any damage caused by the boys' pranks seemed to close the matter with good will on the part of all concerned.[2]

2 Binty Hayward Lloyd, daughter of the owner of the Fox, told the author in 1969 that the villagers thought the Americans good laddies who worked hard and played hard, and they admired them "even though they dumped salt in their ale and mounted their bicycles from the wrong side."

Rosser tended to be impatient with some of the pranks of his truck drivers and confided this several times in his diary, as: "I have trouble keeping the

Except for the fatal Douthit accident and the mean fracture of Webster's arm that refused to heal properly, the boys remained surprisingly healthy. The Mansfield general hospital had given medical assistance when needed. With the exception of a few sore throats and mashed fingers, there had been only three or four occasions when someone was retained in the hospital for a day or so. Fortunately, there had been no serious or extended illness. At the moment Walker could report all the crews healthy and working.

Of course, the biggest factor behind the fine conduct and work record of the boys was the additional food rations now being received from the army depot G-20 at Burton-on-Trent. Even now, the Christmas menu at Kelham Hall would feature the traditional turkey dinner with the open house and the Christmas eggnog. The army depot could supply everything necessary except eggs for the eggnog. Whenever Walker was asked about the egg situation, he always used the occasion to relieve the questioner of all the soap he might have on hand. For several weeks Walker and a few of the boys had been making mysterious calls at the homes of their friends in the Eakring community.

Acting on the old adage that "Cleanliness is next to Godliness," Walker and the boys were finding that some of the housewives were willing to wink at the rationing regulations by making a few innocent trades of their egg rations for the soap rations of the boys. With the help of Mrs. Drummond

---

small K-7 trucks slowed down. Sikes and Norberg try to move everything at one time and see who can make the most trips. The natives hear them coming down the road and jump off their bicycles and take to the ditch cussing the *Bloody* American *Lorries*. I have to raise hell with them every few days.

Walker wrote Ed Holt in July: "The English reaction to us in the community and on the job is that we can get the job done, that we are a noisy, stormy lot, much on the 'Bull in the China Shop' type. The hair-lipped steward in the company canteen expressed it very aptly when queried as to whether he wasn't a Yank. He said, 'Yes, sure I'm a Yank, listen to me— g—damn'l' But we are still those amazing 'Yanks.' "

217

Miller of Home Farm that adjoined the monastery grounds, Walker had been able to build up a cache of several dozen eggs in the Miller home. Thus it was that Walker was not surprised to find a large basket of eggs topped off with a bright red bow that appeared at the big double front doors of Kelham Hall on Christmas morning.

The harmony, health, and decorum maintained at Kelham Hall were also due to the skillful management and time-consuming hours spent by both Rosser and Walker with and for the boys' comfort and welfare. Rosser had had a wealth of experience in working with men of the oil fields. Certainly, the oil industry had developed a special breed of men. They filled the ranks of the so-called roughnecks, most of whom could point with pride to families that went back to the early days of Oil Creek, Spindletop, Mexia, Borger, East Texas, Southern Illinois, Midland, Seminole, and a hundred other oil fields where the industry had carried its search for oil. In many cases, the search was by men who had been willing to gamble their last dollar on the hunch that the bonanza would be found by drilling just a few more feet.

At dinner on the 22nd, Walker called attention to the announcement on the big bulletin board concerning the schedule of mealtimes for Christmas Eve, Christmas Day, and the day after.[3] Many of their friends from the company's offices at Burgage Manor and Eakring and in Kelham village had been invited to the open house that was to be held in the recreation lounge and dining hall of the drilling crews. As an afterthought, Walker cautioned that the fathers and brothers of the monastery had also been invited. Therefore, he said, every man would be expected to conduct himself accordingly.

All were happy and expectant concerning the Christmas festivities, but the clandestine trading of soap rations for eggs remained a mystery. The facts did not become known until

[3] See *Schedule for Christmas Holidays, E* file.

many years later—and then only as a confidential admission by some of the English folk who still live in this beautiful section of Great Britain. It should be said that all concerned in these barter actions felt fully justified in participating in the underground black market affair.

One of the traditional facets of an English Christmas is the Tom and Jerry and eggnog bowl. Eggnog is embedded in the Yule ritual almost as deeply as plum pudding and the custom of singing the well-known Christmas carols.

Christmas, 1943, and the dozens of stories that have since been told and retold by the American oil-field workers to their friends and families, and now to the grandchildren who gather in their homes for celebration of the festive Christmas season, will no doubt become a legend that will live on. One of the boys recently said as he reminisced about the happenings of that day, "All other Christmases will be just another anniversary of the Christmas that came to Kelham Hall in 1943."

Many of the people living today in the Kelham area, and particularly some of the residents of Kelham monastery at the time, have poignant memories of the day. Father Gregory and several of the other monastery personnel failed to recognize the predominant amount of bourbon that had gone into the eggnog recipe until it was almost too late. The liquor was skillfully and pleasantly blended with other ingredients and filled Mrs. Miller's largest punch bowl, which was invitingly perched atop a card table in the recreation lounge. A generous serving of the Christmas dinner and a short walk in the freezing rain that came to Kelham that day fortunately rescued the pious dignity of the men of the church.

A carton of American cigarettes for each man was a gift from Frank Porter that was cherished by all. The pound note that Walker was able to pass out to all of the boys was a token of best wishes from the Noble Company folks back home in Oklahoma. And then there was the great hilarity that came

219

with the elaborate ceremony of presenting Phil Albritton with an iron Victoria cross. Complete with a length of rotary spinning chain, the twenty-pound decoration was hung around his neck in honor of the many little gripes that Albritton was wont to express about certain English practices and a few of the war inconveniences. The Victoria cross with its sizable chain necklace had been carefully, with almost loving deliberation, forged in the D'Arcy workshop at Kelham village.

George and Doris Newton and other couples who had become friends with these young Americans were on hand to help with the serving and to see that everyone was adequately supplied with the innocent-looking, but delicious-tasting, eggnog. Doris Newton remembers taking a tray of used mugs to the kitchen for washing where she and several other of her lady friends were helping out. It was discovered that all of the mugs had not been emptied. The ladies, so completely indoctrinated in wartime practices, did not want any of the luxuries they were enjoying to be wasted. They decided it would be better to consume the eggnog left in the mugs rather than throw it into the waste. Doris says that she is now sure all of them got home safely. But, in her particular case, it being rainy, cold, and muddy, her husband gallantly placed her on the handlebars of his bicycle and miraculously peddled her home without dumping her a single time in the puddles along the roadside. But the nicest thing that she recalls about getting home was George's assistance in getting her up the stairway and into the big four-poster bed which had been handed down through generations of her very proper British ancestors.

The schedule for Christmas at Kelham Hall, and in fact throughout the British Isles in 1943, was to stretch over a four-day period. Christmas Day dinner had been served in two shifts so that all the crews had the benefit of plenty of time to put away the delicious hot food. All of the rigs had been able to shut down at noon at the end of the morning tour of

December 24 except the crews on the Nocton number 3 location.

Nocton number 3 continued to be the big headache. The yellowed pages of Rosser's diary sound the disappointing note: "I roughnecked all last night on the fishing job at Nocton number 3. Nearly everyone is off three days for Christmas. It is pretty hard to make the crews on this well work when everyone else is enjoying the holidays, but the company asked me to do it, so we are really after it. The hole is in a hell of a shape. We're still fighting the damned right- and left-handed drill pipe still in the hole. No matter which way you turn it, it is the wrong way. Sams says if we could find a cross-eyed Chinaman, he might be able to unravel this puzzle. But anyway, all of us feel at home fishing on Christmas. All of us have had to do this before on holidays. One of the boys remembers he had a fishing job on his wife's birthday."

On Sunday, December 26, Rosser hurriedly scribbled: "Today is still a holiday for everyone. It is Boxing Day. I never heard of it before, but after Christmas is another English holiday when everyone opens the Christmas boxes and packages. Since Boxing Day came on Sunday this year, tomorrow will be celebrated as a holiday." Rosser also noted: "I was by Kelham Hall this afternoon; had to go to the machine shop at Eakring. Don reports a good time was had by all. From the way he looked, I think he even got tight. I didn't tell him I had burned out the rods in his car and I'm having transportation problems. Will tell Don later when he is in better shape to hear the bad news."

Rosser's entry the next day was: "Today is still another holiday in jolly old England—Boxing Day, continued. The English are all home opening up their presents. The crew and I are still fishing at Nocton number 3. We are getting the left-hand pipe and overshot out O.K. We got hold of the four-and-a-half-inch drill pipe, but could not pull it."

Finally, on Tuesday, December 28th, Rosser's diary contains the following entry: "Mr. Bremner telephoned the rig this A.M. and told me to plug Nocton number 3 and skid the rig to a new location. I think this is the best thing to do. We could go ahead and clean out the hole, but it would take time and our fishing tools over here are few and sorry."

Thus it was that the story of the American drilling crews' Christmas holidays at Kelham Hall came to a close. The drilling crews on the Nocton number 3 location, as well as Rosser and Sams, were greatly relieved to take another hitch on their tour of drilling duty at Nocton. By the time the morning crew broke tour at noon on Wednesday the rig had been skidded to a new location. Rosser and McGill made a quick trip the next day to Sheffield to get oil filters and grease gun. The weather continued bad. Freezing rain and fog made the roads hazardous and particularly so when a driver could use only the regulation dimmed headlights.

Rosser, however, enjoyed the comforts of a good bed at the Saracen's Head Hotel that night in near-by Lincoln, where he fortunately met Major McDill and Colonel Uexler from the army depot at Burton-on-Trent. The next day Rosser was feeling better. He was able to report that the boys had their fourth rig assembled and would start changing over from the A. C. electric motors to the Waukesha diesel engines the next day.

Two unforeseen events, however, delayed the changeover. Some of the boys had tried to thaw out a frozen water line at the new number 3-A Nocton location by building a fire on the line. The dry oat straw that stood in a ten-acre field around the rig caught fire as a result. Rosser's comment was that it took nearly everybody in Nocton to put out the fire, including the bartender at the Nocton Foaming Mug tavern and his wife and daughter.

Difficulties in switching the National 50 rig number 4 con-

Air-raid damage. The Old Palace Yard and the bent sword of
Richard I statue, with Westminster Hall in background. *Courtesy
Imperial War Museum, London.*

Air-raid damage. The Treasury, Whitehall, with ARF squads at work, October, 1940. *Courtesy Imperial War Museum, London.*

Air-raid shelter. *Courtesy Imperial War Museum, London.*

Cockney humor and the "carry on" spirit during the blitz, September, 1940. *Courtesy Imperial War Museum, London.*

Herman Douthit's funeral procession, Kelham Priory grounds.

American military cemetery, Cambridge, where Herman Douthit is buried.

Kelham monastery.

Rosser and Brother Edgar at Kelham monastery.

Oil-field workers Bob Christie, Gerry Griffin, and J. W. Nickle.

De Talt Havely connecting an electric pumping unit.

Pumping unit at Eakring twenty-five years later. In May, 1939, a test well was drilled at Eakring, and this showed traces of oil; a month later oil in commercial quantities was struck. *Courtesy British Petroleum Company.*

A nodding donkey near Duke's Wood, Nottinghamshire, England. *Courtesy British Petroleum Company.*

tinued. The boys could not keep the oil pressure up on the number 1 motor. Despite the fact that all the bearings had been changed, the pressure would not hold at a point as high as needed. But the trouble proved to be only temporary, and the next day the converted rig was operating, Sams reported, "like Grandma's new sewing machine."

John Norberg was visiting his brother Carl and spent the night at Kelham Hall. John was in the air corps and stationed at a nearby base. He was considerably excited about the latrine rumors emanating from his base that the Supreme Command was all set for the invasion of France at a point directly across the channel from Dover. Although the rumor was not to become a fact for nearly six months, it was, nevertheless, exceedingly accurate as to the area where General Eisenhower's armed forces, aided by American and British air and naval support, would in the following June establish three important beachheads in Normandy.

As the holiday period drew to a close, the pressure on the American team to produce more oil eased off. The oil situation in Great Britain had greatly improved since September, and the prime minister directed the minister of war transport to increase evening bus service by twenty-five per cent in the London area the following month. This signaled the beginning of the end to drilling in the Sherwood Forest area.[4] The team looked forward to the coming year and its great events, but looming large in their minds was the thought of embarking for home.

[4] Churchill, *Closing The Ring*, Appendix C, 665, 667.

XVIII.

Preparation for Going Home

Christmas, 1943, was now history at Kelham Hall. New Year's, 1944, was at hand. Competition among the American drilling crews was a natural development. The boys, originally employed as drillers, had continued as head of the crews to which they had been assigned, except in the case of H. A. Hobbs, who had quit to return home and had been replaced by Lewis Dugger. There had been a few intracrew changing of positions for temporary periods because of illness, injury, and brief periods of needed days off for rest and relaxation. Ken Johnson had been moved to a drilling crew when Ray Hileman was given the job as driver on Ole Lo-Go.

The friendly spirit of competition contributed a great deal to the rapid well completions, much to the amusement and satisfaction of the D'Arcy and Anglo-Iranian officials. Southwell had at one time suggested that some sort of prize, or perhaps a bonus, be given to the crew that drilled the greatest amount of footage during the contract period. Rosser knew from experience, however, that too much haste was dangerous so he did not follow up on the suggestion. He believed that if the crews kept a steady pace of normal operations, that accidents, fishing jobs, and other well troubles would be avoided and more footage would be drilled. Since the contract period would soon be drawing to a close, Rosser had, at Noble's suggestion, reviewed the past year's work with Southwell for the purpose of determining D'Arcy's desire with respect to extending the contract. Walker prepared a summary of well completions for each four-week period. This started with April 12, 1943, and continued through January 16, 1944, inclusive, showing the Noble and Fain-Porter crews were entitled to be credited with ninety-four well completions, based on twenty-five hundred feet drilled counting as one well. Seventy-six of the wells were producers and had joined the army of nodding donkeys now in the field. Despite the situation at home, Noble and Fain-Porter companies continued to urge Rosser and Walker to make it clear to D'Arcy and Anglo-Iranian Company officials that the work of the drilling crews would be completed to the exact letter of the contract regardless of any other circumstances. If the boys were agreeable and the D'Arcy and Anglo-Iranian companies desired, the crews would continue their drilling program.[1]

The news from home made the boys aware that the military draft and the maximum production of all kinds of war materials were creating America's greatest of all war shortages, the shortage of man power. The war was being felt at home. They learned of the growing number of former schoolmates and friends who had gone into war plants and those who had been drafted or enlisted in military service. The letters told, also, of those who would someday be coming home in the regulation flag-draped casket.

News from Italy had a sobering effect upon the boys. Several of their friends and former coworkers were members of the Forty-fifth Division which included units of the national guard from Southwestern states, called into service in September, 1940. The Forty-fifth was among the ground forces used January 22 to establish the Anzio beachhead. This was a move to

1 "Noble to Rosser," January 28, 1944, and cablegram, January 26, 1944, *E* file.

free the Italian mainland, but the Germans fought desperately to drive the allied forces into the sea. Casualties were heavy on both sides.

Lloyd Noble wrote from Edmonton, Canada, that he was hurrying back to Oklahoma to attend the funeral of Leslie Fain, the youngest copartner in the English venture. Fain had married in 1926 to Winnie Mae Hall, daughter of Frank C. Hall, Chickasha, Oklahoma. Hall in 1931 sponsored Wiley Post's history-making 'round-the-world' flight in the Lockheed plane called the "Winnie Mae." Fain had loaned his somewhat larger model of the Lockheed 12 to friends in Oklahoma City for an emergency trip to Washington, D.C. They left the Oklahoma City airport at 6:10 A.M., February 4, and amidst sleet and fog in the early evening crashed into Rich Mountain near Elkins, West Virginia. Fain, in Phoenix, Arizona, at the time, was found dead in his bedroom the next morning from a heart attack apparently brought on by the news of the death of five close friends.[2] The tragedy was doubly poignant to Don Walker. He had known Les Fain since the younger man was a freshman at the University of Oklahoma. He had for the past several months been corresponding with him and was expecting another letter from him any day.

Rosser and Walker, as suggested in a letter from Mr. Noble, called the group together to tell them that the 365 days of drilling operations for each of the National 50 rigs, provided by the contract with the D'Arcy Company, would be completed about March 1, or soon thereafter. In order for all parties to make plans accordingly, Rosser explained, it was necessary to

---

[2] Killed in the crash were Roy Hunt, famed speed and acrobatic pilot; his wife, a WASP, who had been ferrying planes for the air force; the copilot George Ruddy; Wilbur E. Hightower, president of the largest financial institution in Oklahoma, the First National Bank and Trust Company of Oklahoma City, and an outstanding civic leader; and his daughter, Mrs. John Roby Penn, III. See the *Daily Oklahoman* (Oklahoma City), February 5, 6, 7, 1944; also "Noble to Walker," February 7, 1944, E file.

know now how many of the boys were willing to stay on in England for continuing drilling operations that the D'Arcy Company might deem desirable. For a moment there was silence while the boys considered this important question. As they began to speak up individually to express their thinking and desires, it quickly became clear that the predominant feeling among them was, that it was now time to return home.

Rosser and Walker immediately got off a cable to the Noble and Fain-Porter companies that only thirteen of the thirty-seven men were willing to remain in Britain on a temporary basis providing they would be home in the near future, or at such time as they might elect to terminate their employment.[3] In view of the small number willing to continue on, Southwell decided the contract work should be closed out as near the first of March as possible. This would require considerable paper work on Walker's part. He would need assistance from Rosser and some of D'Arcy's men. Rosser and Walker cabled home that they had learned that Jackson in New York had arranged with the British minister of transport for the boys to be returned to New York during the first half of March.

The boys would have an opportunity to spend a few days in London, or other places they desired to visit in Britain before being shipped out, as promised by Rosser on several occasions. It was arranged that as each American rig completed 365 days of drilling operations, the crew members would be at liberty to go wherever they pleased. Walker must be kept advised, however, as to how and where they could be reached in the event travel arrangements should be completed during their absence.

General Lee had secured free tickets for the entire group to Irving Berlin's Broadway hit, *This Is the Army*, now showing in London, for the previous 15th of November, but because of Douthit's fatal accident, the trip was cancelled and the tickets

3 "Rosser and Walker to Noble Co.," cable, January 27, 1944, *E* file.

returned to the American Embassy. Now, as each rig completed its scheduled drilling time most of the boys headed for London to see the show that was free to all GIs, a type of entertainment more appealing to them than England's many places of historical significance.

By the spring of 1944 the military successes of the Allies had dimmed the memory of the battle of Britain and the horrors of the London blitz. The enemy was driven from North Africa and the follow-up invasions of Sicily and Italy had opened the Mediterranean. With America's growing defense against the submarine menace in the Atlantic, military equipment and supplies were now piling up to mountain high proportions at military bases throughout Great Britain. Three million Allied fighting forces at British bases awaited the day and hour when the war would be carried across the channel to the enemy-held mainland of Europe.[4] This stepped-up war activity did not cool the determination of the oil-field workers to get home and into the conflict.

Southwell and other D'Arcy officials made it clear that Rosser was needed in England for a few weeks after the drilling crews, accompanied by Walker and Sams, had departed. Although the necessity for staying was plain to Rosser, the fact he would not be going home to his wife and young daughter, whom he had not seen for more than a year, raised a sadness that was difficult to put aside. He was grateful when Lewis Dugger said he would stay with Rosser because young Mrs. Dugger, a commissioned army nurse, had just arrived in England for assignment to duty.

On the week end that the American rigs finished their contract operations of 365 days, Rosser and Dugger took off for London for the purpose of celebrating Mrs. Dugger's arrival. Wartime youth needed no encouragement for celebrations,

[4] Sulzberger, *World War II*, 280.

but an excuse was always welcome. Friends at the American Embassy, as usual, had been happy to make reservations for them wherever rooms were available in the now crowded London hotels.

Rosser, perhaps with a shade of envy at the Duggers' reunion, was careful to see that they were provided with a suite at the Hotel Carlson. He got a room in one of the smaller hotels in a side street near Picadilly Circus. No one could have foreseen that Hitler's air force would make a final desperate retaliatory attack on London that night. That Saturday night of March 4, 1944, will always be remembered by the Duggers and Rosser as the night they came face to face with World War II. The following morning Rosser was out early. He was appalled by the extensive damage done to the center of London. His attempt to telephone the Duggers got only the operator's report that the Carlson's number was out of order.

Rosser made his way with some difficulty to the Carlson since the air raid wardens had cordoned off walkways to avoid dangers from weakened buildings. At last he reached the block in which the hotel stood. His heart skipped a beat and his spirits dropped. The Carlson had been hit! One wing lay in ruins! A man who appeared to be part of the management of the hotel told him the Duggers were not in the wing of the hotel that had been hit and they had been evacuated in the night.

Headlines of the papers now on the streets proclaimed the terrible news that over 600 Londoners had been killed outright and more than a thousand critically injured. Hospitals were continuing to receive injured victims. A warden directed Rosser to a Red Cross check point where an incomplete alphabetical list of the dead and injured was being compiled. The Red Cross lady in charge was helpful. She checked her own list as well as other Red Cross stations that had telephone connections. Finally, she assured Rosser that the name Dugger did not appear on any of the killed or injured lists she was able to check.

With nothing to do but wait, Rosser returned to his hotel. He hoped that the Duggers would call as soon as possible, but the waiting time seemed like an eternity. At last, near noon, the call came through from a village in Wales. The Duggers were safe and would take a train direct to Newark-on-Trent. Rosser and the Duggers arrived in Newark in the early morning hours of Monday, March 6. The three had breakfast at the Clinton Arms Hotel and retold the experiences of the horrible night. Dugger laughingly says he had been married long enough to realize Mrs. Dugger was the boss, but he had no idea Hitler was so afraid of her arrival in Great Britain.

Rosser and the Duggers did not know that the Luftwaffe's bombing raid over London was a retaliatory strike ordered by Hitler over the protest of his top generals who were attempting to persuade him to concentrate on the production and use of the V-1 and V-2 rockets which the German high command and the Allies, as well, knew to be the most destructive of all their weapons of war. The news stories and radio reports coming through to Kelham Hall had mentioned the heavy day-and-night bombing raids over German-held western Europe since the first of the year. General Dwight D. Eisenhower, Supreme Commander of the Allied Forces in Britain, had already received the explicit order on February 12:[5] "You will enter the continent of Europe and in conjunction with the other Allied nations, undertake operations aimed at the heart of Germany and the destruction of her Armed Forces."

As *Time* magazine put it:[6]

The timetable was set, the weapon almost forged, the London press reported "millions of U.S. soldiers pouring into Britain ("straphanging across the Atlantic," one newspaper called it).

[5] Eisenhower, *Crusade in Europe*, 225.
[6] *Time Magazine Capsule 1944*, 69, 70.

Now the men to command the attack on the German heartland were chosen. The division of command gave sobering impressions: in the hard, costly attack from the west, U.S. troops will bear a major burden.

General Dwight David Eisenhower will direct the main assault from Britain. On the supple, affable shoulders of the 53-year-old American will fall the toughest job of military coordination since Marshal Ferdinand Foch took supreme command over 1918's Western Front.

"Ike" Eisenhower had to defend his conduct of the Italian campaign last week. But Messrs. Roosevelt & Churchill had their reasons for assigning him to the west. He is the only commander with experience in large-scale coordination of British and American forces. He is particularly respected by the British; Winston Churchill has called him "one of the finest men I ever knew." He has more schooling than any other commander in the management of amphibious operations (French North Africa, Sicily, southern Italy) .

Sixty thousand men and women labored through a cold winter's day in Britain last week. Seven thousand R. A. F. men listened to briefings. Then from airdromes all over England, 1,000 monster Lancasters and Halifaxes sped for Berlin.

Berlin's antiaircraft cannon hammered with fury and 43 of the four-engined bombers were downed. But within the space of half an hour 2,800 tons of explosives plummeted into the torn city— 90 tons a minute. As the armada headed home, smoke from the fires of Germany's capital rose 20,000 feet in the air. It was the war's heaviest raid on bleeding Berlin or anywhere else.

XIX.

Farewell

The contract between the Anglo-Iranian Company and the Noble and Fain-Porter companies had been terminated. "Present plans which Southwell approves," Rosser cabled. "All arriving New York around March 20th. Suggest you advise families."

The magnificent drilling record made by the Noble and Fain-Porter companies' drilling crews in England had resulted in credit for 106 wells measured by one well for each 2,500 feet drilled. Ninety-four of the holes drilled had been completed as oil producers. The Nocton location well number 3, an outside test well, was the only well plugged and abandoned as a lost hole by the American crews. This was a remarkable record that remains an amazing feat under the circumstances of wartime shortages and hardships prevailing in Great Britain during 1943 and the early part of 1944. Not counting the men who left the group to return home before completion of the contract, a total of thirty-seven men remained.

It was, therefore, no surprise to Walker when Southwell advised him on the morning of March 2, that passage had been arranged on the passenger vessel HMS *Mauritania*, also serving the Allied forces as a troop transport. The *Mauritania*, he said, was scheduled to sail at 6:00 P.M. on the following day, March 3, 1944, from Greenock docks at Prestwick, the port for the city of Glasgow, located approximately three miles beyond the western edge of the city.

Walker had the immediate job of corralling the thirty-seven youthful roughnecks who had been turned loose on Great Britain. The notes left by each man as to where he could be contacted revealed the group was scattered far and wide. Nevertheless, Walker commenced the task of telephoning all whom he could reach and requested them to tell all others they could find, to return to Kelham Hall by 3:00 P.M. of the following day. The group with their luggage would be transported to Newark-on-Trent to board the 4:00 P.M. train to Glasgow.

The entire area had been hit by heavy snows on two successive nights and travel in the Kelham, Eakring, and Southwell areas was slowed almost to a standstill. With the information that had been left, Walker was surprised to find so many of the boys could be reached so quickly by telephone. The calls continued through the day and night of March 2. By the following morning, everyone had been notified except Jim Harding, who had failed for some reason to leave information on where he could be contacted. Rosser had gone to Edinburgh to settle a number of insurance claims the boys had pending with the Caledonia Insurance Company. When reached by Walker, Rosser said that he would call Thomas Cook & Sons in London to have a freight car attached to the train for the boys' luggage which should be at Newark before train time on the following day. He would also call the transportation man, Sergeant Warren, at the American Embassy and request him to give the Cook company such assistance as he could to get the freight car.

Walker had some anxious moments about the return of several of the boys who were at a "wildcat location" near Southport, about forty miles from Kelham, helping the British boys move one of the D'Arcy rigs on what proved to be the last outside test location to be drilled. The heavy, drifted snow made car travel to Southport extremely difficult. Although slow, the trip was made successfully with Robbie Robinson driving. The boys arrived at Kelham in time to pack and join the others. In the meantime, Rosser called to say he would meet them at the Glasgow station where a bus and a truck would transfer them and their baggage to the Greenock docks.

Packing and making other preparations to leave in time to board the four o'clock train at Newark resulted in the problem of the disposal of work clothes, gloves, hard hats, hard-toed safety shoes, "long-handle" underwear; the latter had been a subject of mirthful gossip among the cleaning and washwomen of the village. Considerable confusion was caused by the wrapping and packaging of the clocks that Morton had acquired for his collection. Although he is now the owner of a valuable collection of rare clocks, those he brought home with him still take first place in his sentimental feelings for the collection.

A real holiday spirit pervaded the monastery and for the second time during the boys' sojourn in Great Britain, the monks opened the monastery doors to the public, including the ladies of the D'Arcy Company office in Burgage Manor and people of Kelham village and the neighborhood, so that all the friends of the Americans could say farewell before they left. The boys' ingenuity was displayed to the last. Someone started an auction to sell the unwanted clothing and equipment. Some of the gaudy shirts brought to England with them went to the English lads for a few shillings. Work clothes, in most cases, were given to English friends who were happy to get the useful garments.

At last all were packed and ready to go except Walker and Harding. Walker arrived rather breathless shortly before four o'clock to announce that the railroad people had telephoned to say the train would not arrive in Newark until about 8:00 P.M. This welcome bit of news gave the group more time to get rid of their clothing and say farewell. Walker and Watson visited their good friends, the Millers, at their nearby farm. Mrs. Miller remembers the occasion with much sentiment. She insisted that Walker and Watson have dinner before leaving, but a bit of bacon and powdered eggs was the only wartime fare she was able to furnish for the occasion. But this mattered little. The big loss to the Millers and the many others of Kelham village was that their American friends were going home.

The D'Arcy Company had helpfully arranged for the bus that would pick up the boys and take them to the railway station. Still, the one man, Harding, was missing. But just as the baggage was being loaded, he mysteriously showed up, greatly to the relief of the rest of the party. Walker remembers that several of the boys had packed for him, but the new custom-made suit and topcoat he had bought in London were worn for the occasion.

Packing completed, baggage sent ahead, fond farewells said, the boys were then delivered to the railway station in Newark. All were aboard when the train pulled out for Glasgow. It was past midnight when they clamored out of the train at the Glasgow station to meet Rosser with the truck and bus standing by. Rosser explained they must hurry with loading and getting to the docks. The captain of the *Mauritania* was raising Billy hell and was fit to be tied. The delay in sailing was causing the *Mauritania* to be charged $140.00 an hour demurrage for her failure to clear the docks on time.

Walker had picked up the berthing tickets for the boys and had hurriedly taken their passports from the D'Arcy office safe at Burgage Manor. He held the large folder containing these valuables in his hand with a tight grip.

It was cold, clammy, and dark in Glasgow. The lights of the train that pulled to a stop in the railway station were already dimmed. The outside lights of the station appeared to be smothered in the heavy night fog that lay close to the ground. The confusion of the men in checking their baggage to make sure it was put on the truck for transfer to the dock was tremendous.

Rosser had said his good-bys to the gang and assured everybody that he would be seeing them in the very near future. He decided not to add to the confusion by accompanying the group to the docks, since Walker seemed to have everything under control. At the moment, Rosser and Walker stood aside awaiting word from the driver that he was ready to start. The awkward moment of silence between the two men seemed to encourage each to express a farewell gratitude for the support the other had given through the many experiences they had shared over the past year. Suddenly, both men were attempting to talk at once. The tight clasp of the handshake had come with the call of the driver. It released a simple expression of the warm and lasting relationship that had been kindled between these men so extremely opposite in nature, characteristics, and backgrounds. But both recognized, in this hurried moment of leavetaking, that they had the common denominator of respect and loyalty for each other that would be the genesis of a deep and lasting friendship. The noise of the racing motor bespoke the bus driver's impatience.

As Walker climbed aboard and the bus pulled away in the darkness, Rosser stood for a long moment in the cold of the night watching the boys wave at the windows. Snow had started falling again. He noticed how the large flakes glistened in the dim station lights as he pulled his overcoat tighter around his neck and plunged both hands deep into the warm side pockets. Then remembering his appointment at the D'Arcy Company offices the next day to deal with details relating to the completion of the drilling contract, he started moving in the direction of the station to check on the departure of the next train for Newark-on-Trent.

A baggageman pulling a small handtruck stopped to stare at Rosser when the latter suddenly discovered his passport was missing and stamped a foot in the soft snow. He yelled into the cold and empty night, "That little bastard Walker has left with my passport!" Still fresh in his mind today is the thought he had that night. "What the hell am I going to do in England while my passport is with Walker on the high seas?" He stood stock still as the train hove into sight, with his question unanswered.

Aboard the train, Rosser dropped into the corner seat of his compartment for a nap on the run to Newark. But sleep or rest was not easy to come by for Rosser at this time. The colorful parade of the amazing events of an unbelievable year crowded into his memories. The big question he had often asked himself in the last few hours was: had the expenditure of money and the efforts of the many people involved been justified by results obtained by the Noble and Fain-Porter men? Rosser was convinced that he could objectively answer yes to the question. A number of reasons confirmed his answer.

First, production from the oil fields of the English Midlands had risen during the year 1943 to a peak daily production of slightly more than 3,000 U.S. barrels of oil per day. By the end of 1943, Eakring-Duke's Wood and Formby had sent 2,289,207 U.S. barrels of the high grade paraffin base oil to refineries on the west coast and in the south of Scotland. Although Rosser could not know at the time, 1944–45 would add another 1,231,346 to the total U.S. barrels shipped to refineries, making a total of 3,520,553 U.S. barrels produced and moved to refineries from Great Britain's own oil fields by the end of 1945.

Rosser took pride in the fact that his men had met the challenge desperately urged by Britain in 1942 and had come to her aid when oil stocks faced depletion. Their American know-how and equipment tapped the surest source of replenishment when every barrel of oil helped ease the shortage. He recalled that morning when a record move from one location to another was made in six and one-half hours, and the morning when an outfit was shut down, transferred to a new site, set up and drilled 960 feet, all within twenty-four hours. Now, his team was disbanded, on the way home, some bound for military service. He wondered how many would make their way back to Britain, next time in uniform. It appeared to him the United Kingdom was bulging with airfields, storage sites, hospitals, and mobilization camps. From what he had seen and heard, he guessed these would be too late for the first assault on Fortress Europe. But, he felt, they had already made a contribution.

All benefits flowing from the United Kingdom project probably could not be tagged with a dollar value, but Rosser felt the officials of the D'Arcy and Anglo-Iranian companies would agree with him that the many practices their men had learned and adopted from the American drilling crews would continue to be of great value to them in future years.

No longer would the drilling be shut down for a fixed seventy-two-hour waiting period for cement to set when bottom-hole pressures, temperatures, and water chemicals indicated a shorter period would be sufficient.

No longer would drilling wells in areas where high reservoir pressures did not exist labor through heavy drilling muds to bring the bit cuttings to the surface. Now they know that fresh water would do the job much easier and faster, thus increasing the speed of penetration of the drill bit.

No longer would the drilling teams, as Southwell and Sir William Fraser still insisted on calling the crews, be required to wait until they needed additives for drilling mixtures or cement, before it was delivered at the drilling location. Hereafter, these additives and cement would be delivered well in advance of their need.

No longer would time be lost in pulling the drill stem for replacement of the bit on a fixed hourly schedule. Hereafter, the bit would be left on the bottom of the hole as long as it was satisfactorily penetrating the geological formation.

No longer would time, labor, and expense be lost in tearing down and rebuilding old-style wooden derricks. Now the unitized draw-works equipped with jackknife mast could be lowered and raised in a matter of a few hours for skidding the rigs to a new location.

No longer would a large number of men be required to dismantle and rig up drilling equipment for moving the drilling rig from one location to another.

No longer would time and human effort be spent in loading heavy equipment onto trucks. Hereafter, the magic of self-loading trucks with winches and gin poles would do the work with only the labor of attaching the lines to the equipment to be moved and pushing a button on the dashboard of the truck.

Rosser also felt good about the thirty-six D'Arcy field men who worked on the rigs to learn the ways of the American drillers, who he was sure would go on multiplying their beneficial knowledge.

Then, of course, as Rosser told himself, there was the great intangible value symbolized by the nodding donkeys of Sherwood Forest of which Noble and Porter would be most proud. This was the fine relationship their men had developed with the people of Great Britain and the understanding and knowledge the Americans had brought to the English Midlands.

Bibliography

Manuscript Collections

E file. A confidential file (the E standing for England) which was our main source of information. It is an extensive collection of private letters, drilling reports, contracts, official documents, telegrams, cablegrams, reports of employees' injuries, insurance claims, and a complete personal history for each of the forty-four American oil-field workers who, in 1943–44, drilled for oil in Sherwood Forest, Nottinghamshire, England. To prevent leakage of information, less than a dozen personnel of the Noble and Fain-Porter drilling companies had access to the file. Release of restrictions on *E* file was permitted to the author in 1968. The file is now in possession of the Noble Drilling Corporation, Tulsa, Oklahoma.

George Otey file, less voluminous than *E* file, contains legal documents, correspondence between the officials of the Anglo-Iranian Co., a branch situated in New York, and George Otey who acted as legal adviser to both Noble and Fain-Porter in their relation to the English drilling project. In chapter notes, the Otey file will be referred to as *O* file. It is in possession of Mr. George Otey, Sr., Ardmore, Oklahoma.

Miscellaneous

American Foreign Service Report of the Death of an American Citizen. London, England, December 4, 1943.

Certified Copy of an Entry of Death. Entry 104 in the Register Book of Deaths for the Sub-District of Sutton-in-Ashfield in Notts Co. England, November 19, 1943.

Contract between Anglo-Iranian and Noble-Fain Porter Drilling Companies.

Contract between Fain-Porter and Noble Drilling.

Employment Work Agreement.

Eugene Preston Rosser's diary for 1942–43, a carefully kept, day-by-day account of Rosser's experiences and the experiences of his men in 1942–43 when dispatched to England for a year to drill for oil in British oil fields. Diary at present in custody of E. P. Rosser.

Information of Witnesses, severally taken and acknowledged on behalf of our sovereign Lord the King, touching the death of Herman Douthit at the 30th Gen. Hospital, Sutton-in-Ashfield on the 16th day of November . . . .

Notes on Kelham and Aversham in Relation to Families of Sutton and Manners. Made by Antony Snell and T. Jones in 1939 in Archives of SSM monastery at Kelham, Nottinghamshire, England.

Scrapbook. Donald E. Walker's scrapbook containing personal notes from British friends, theater programs, menus, and Anglican church services attended by Walker when living in an Anglican monastery in 1943 (together with Rosser, the superintendent of secret drilling operations in the U.K., and forty-two American oil field workers) has been consulted by the author.

Walker's two small black notebooks. Kept in 1943 by Donald E. Walker. These are intimate references as to the nature and quantities of food purchased for oil-field workers, the debts they owed to fellow poker players, cleaning bills, and other miscellaneous personal matters.

Documentary Film

Documentary Film of Drilling Operations in Sherwood Forest in 1943. Courtesy Anglo-Iranian Oil Company, now the British Petroleum Company.

Tapes

Tapes by Eugene P. Rosser and Donald E. Walker made in 1968, 1969, and 1970.

Personal Interviews

Mr. R. S. Adams, Rose Cottage, Eakring, Notts, England, 1969.

Mr. C. M. Adcock, Elysium, Whiteleaf, 1969.

Prince Risborough Bucks, England, 1969.

Father Ernest Ball, S.S.M., Kelham, Newark, Notts, England, 1969.

Mrs. Nancy Henson Bissel, Kelham, Newark, Notts, England, 1969.
Mr. Lewis Dugger, Tulsa, Oklahoma, 1971.
Mrs. Binty Hayward, Loughboro, Notts, England, 1969.
Mrs. Drummond Miller, Home Farm, Kelham, Newark, Notts, England, 1969.
Mr. George Montgomery, Eakring, Notts, England, 1969.
Mr. and Mrs. George Newton, Kelham, Newark, Notts, England, 1969.
Mr. George Otey, Ardmore, Oklahoma, 1971.
Mrs. Lucy Richardson, Kelham, Newark, Notts, England, 1969.
Mrs. Audrie Ross, B. P. Co. Office, Eakring, England, 1969.
Mr. Eugene P. Rosser, Houston, Texas, 1965–71.
Mr. Gordon Sams, Tulsa, Oklahoma, 1970.
Wallace and Elsie Soles, Southwell, Notts, England, 1969.
Sir Philip Southwell, Manor House, Tendring, Essex, England, 1969.
Mr. Donald Walker, Tulsa, Oklahoma, 1967–71.
Father Gregory Wilkins, House of the Sacred Mission, Kelham, Newark, Notts, England, 1969.

Master's Thesis

Bradley, Francis T. "The British Commonwealth and Its Petroleum." University of Tulsa, 1954.

Books

Adamic, Louis. *Dinner at the White House.* New York, Harper & Brothers, 1946.
Ambrose, Stephen E. *The Supreme Commander: The War Years of General D. Eisenhower.* New York, Doubleday, 1969.
Bishop, Edward. *Their Finest Hour: The Story of the Battle of Britain 1940.* New York, Ballantine Books, 1968.
Blumenson, Martin. *Sicily: Whose Victory?* Ballantine's Illustrated Campaign History of World War II, Book No. 3, New York, Ballantine Books, 1969.
Boussel, Patrice. *D-Day Beaches Revisited.* New York, Doubleday, 1965.
Bradford, Ernie. *Wall of Empire: The English Channel.* New Jersey, A. S. Barnes, 1966.
Bradley, Omar N. *A Soldier's Story.* New York, Henry Holt, 1951.
Broad, Lewis. *Winston Churchill: The Years of Achievement.* New York, Hawthorne Book Publishers, 1963.
Butcher, Captain Harry C., USNR. *My Three Years with Eisen-*

hower. New York, Simon and Schuster, 1946.

Churchill, Winston S. *Blood, Sweat, and Tears.* New York, G. P. Putnam's Sons, 1941.

————. *Closing the Ring.* Boston, Houghton Mifflin, 1945.

————. *The Gathering Storm.* Boston, Houghton Mifflin, 1948.

————. *The Hinge of Fate.* Boston, Houghton Mifflin, 1950.

————. *The Grand Alliance.* Boston, Houghton Mifflin, 1950.

————. *Their Finest Hour.* Boston, Houghton Mifflin, 1949.

Curtin, Thomas. *Men, Oil and War.* Chicago, Petroleum Industry Commission Publishers, 1948.

Dempster, Derek. *The Narrow Margin.* New York, Coronet Communications, Paperback Library Edition, 1969.

Editors of *Look* Magazine. *Oil for Victory: The Story of Petroleum in War and Peace.* New York, Whittlesey House, 1946.

Eisenhower, Dwight D. *Crusade in Europe.* New York, Doubleday, 1940.

Farago, Ladislas. *Patton: Ordeal and Triumph.* New York, Arthur Barker Publisher, 1964.

Fitzgibbon, Constantine. *London's Burning.* New York, Ballantine Books, 1970.

Frey, John W., and H. Chandler Ide, eds. *A History of the Petroleum Administration for War: 1941–1945.* Washington, D.C., Government Printing Office, 1946.

Goodhart, Philip. *Fifty Ships That Saved the World.* New York, Modern Literary Editions Pub. Co., 1965.

Hechler, Ken. *The Bridge at Remagen.* New York, Ballantine Books, 1957.

Ickes, Harold L. *Fightin' Oil.* New York, Alfred A. Knopf, 1943.

Keegan, John. *Waffen S. S.: The Asphalt Soldiers.* New York, Ballantine Books, 1970.

Killen, John. *A History of the Luftwaffe (1915–1945).* New York, Berkley Publishing Corp., 1967.

Kieran, John, ed. *Information Please Almanac 1949.* New York, Farrar, Straus and Company, 1949.

Klein, Alexander. *The Counterfeit Traitor.* New York, Henry Holt, 1958.

Langer, William L., ed. *An Encyclopedia of World History.* Rev. ed. Boston, Houghton Mifflin, 1948.

Longhurst, Henry. *Adventure in Oil: The Story of British Petroleum.* London, William Clowes and Sons, Ltd., 1959.

Martin, Ralph G. *The G. I. War, 1941–1945.* New York, Avon Books, 1968.

Mosley, Leonard. *On Borrowed Time.* New York, Random House, 1969.

Popple, Charles Sterling. *Standard Oil Company (New Jersey) in World War II.* New York, Standard Oil Company (N. J.) Publishers, 1952.

Pyle, Howard. *The Merry Adventures of Robin Hood.* Racine, Wisconsin, Whitman Pub. Co., 1955.

ROTCM 145–20: Department of the Army ROTC Manual. American Military History 1607–1953, Washington, D.C., Dept. of The Army, July, 1966.

Rothburg, Abraham. *History of World War II.* New York, Toronto, London, Bantam Books, 1962.

Ryan, Cornelius. *The Longest Day.* New York, Fawcett World Library, 1960.

The Science of Petroleum: The World's Oilfields. Vol. 6, Part 1. London, Oxford University Press, 1953.

Semmes, Harry H. *Portrait of Patton.* New York, Coronet Communications, Paperback Library, 1970.

Sulzberger, C. L. *World War II in Europe 1939–41.* New York, American Heritage Press, 1970.

Time Magazine Capsule 1940, New York, Time-Life Books, 1968.

Time Magazine Capsule 1941, New York, Time-Life Books, 1967.

Time Magazine Capsule 1943, New York, Time-Life Books, 1968.

Time Magazine Capsule 1944, New York, Time-Life Books, 1967.

Tuchman, Barbara W. *Stilwell and the American Experience in China 1911–45.* New York, Macmillan Company, 1971.

Vader, John. *Spitfire.* New York, Ballantine Books, 1969.

Warren, C. E. T., and James Benson. *The Midget Raiders.* New York, Ballantine Books, 1968.

Whiting, Charles. *Decision at St.-Vith.* New York, Ballantine Books, 1969.

———. *Bradley.* New York, Ballantine Books, 1971.

———. *Patton.* New York, Ballantine Books, 1971.

Who's Who 1971–72. New York, St. Martins Press, 1971.

Wilmot, Chester. *The Struggle for Europe.* New York, Harper & Brothers, 1952.

World Almanac 1971. Luman H. Long, ed. New York, Newspaper Enterprises, 1971.

Young, Peter. *Commando.* New York, Ballantine Books, 1969.

Articles

Adcock, C. M. "Three Decades of Drilling," *BP Shield,* London (March, 1968), 20–21.

"Baptism of Fire," *The Lamp,* Vol. XXIV, No. 5 (February, 1942), 16–19.

"The Cargo Plane's Priority Passenger," *The Lamp,* Vol. XXV, No. 5 (February, 1943), 6–11.

Day, Alec H. "Continuing Search for Oil in Britain," *World Petroleum,* Vol. XVII, No. 10 (September, 1946), 47–49.

Dickie, R. K., and C. M. Adcock. "Oil Production in the Nottinghamshire Oil Fields," *Journal of the Institute of Petroleum,* Vol. XL, Pt. 2 (July-December, 1954), 188.

"Four Heroes," *The Lamp,* Vol. XXV, No. 4 (December, 1942), 3.

Holman, Eugene. "Oil Transportation in a World at War," *The Lamp,* Vol. XXIII, No. 5 (February, 1941), 6–10.

"How the Germans Lost the War of Oil," *The Lamp,* Vol. XXVII, No. 3 (June, 1945), 1–7.

"Lost in Clipper Crash," *The Lamp,* Vol. XXV, No. 5 (February, 1943), 4.

MacCormac, John. "Britain Hits Oil, 100,000 Tons a Year," *The New York Times* (September 25, 1944).

Mantach, Joseph T. "Business As Usual in London Town," *The Lamp,* Vol. XXIII, No. 6 (April, 1941), 8–12.

"Oil in the British Isles," *The Lamp,* Vol. XXI, No. 3 (October, 1938), 21.

"Private Bill Jones, and the Rest of Us," *The Lamp,* Vol. XXV, No. 6 (April, 1943), 1–3.

Southwell, C. A. P. "Petroleum in England," *Journal of the Institute of Petroleum,* Vol. XXXI, No. 254 (February, 1945), 27–39.

"The Story of Jersey's Part in the War; Technical and Engineering Work Carried on for Many Years Suddenly Assumes Telling Value in Arming the United States and Its Allies," *The Lamp,* Vol. XXIV, No. 6 (April, 1942), 5–9.

Wilmoth, V. J. "Some Notable Wartime Oil Fires," *Journal of the Institute of Petroleum,* Vol. XXXII, No. 265 (January-December, London, 1946), 1–18.

Pocket Editions, Reprints, Pamphlets

Appleby, R. V. *A History of Newark-On-Trent.* Printed and pub-

lished by J. Stennett, Newark-on-Trent, n.d.
A Concise Guide to Southwell and the Rural District. Southwell, Notts, England, n.d.
Gregory, Father, S.S.M. *Kelham.* n.p., n.d., 15–17.
Heywood, Hugh, provost. *Southwell Minster.* 8th ed. Southwell, England, 1968.
In the Shade of Sherwood Forest. [A pamphlet presumably published by B.P. Ltd.] n.d.
Our Wonderful Empire. Birn Brothers, Ltd., London, n.d.
Smith, Turner (Sparky). *My Guide in Boots and Stetson.* n.p., n.d.
Southwell, C. A. P., M.C., B.Sc. *Developing the English Oilfields.* n.p., n.d.
Thoresby Hall: An Illustrated Survey and Guide. Designed and produced by Eng. Life Publications, Ltd., Derby, England, 1967.

Newspapers

"Capitol Demands Unified Command of Armed Forces," *Tulsa Daily World* (January 26, 1942). [Vol. no. and page missing.]
Clark, Lee. "D-Day Talk Is Common," *Tulsa Daily World* (March 5, 1944). [Illustrated by map of European coast.]
"England's Oil Fields," *Daily Oklahoman* (Sunday, October 22, 1944).
"Farewell to Lindbergh" (an editorial). *Tulsa Daily World* (October 8, 1941), 11.
Front page, Magazine Section, *Tulsa Daily World*, Tulsa, Oklahoma (Sunday, June 14, 1942). The American flag in color accompanied by this quotation from Gen. Douglas MacArthur: "Take every other normal precaution for protection of the headquarters, but let's keep the flag flying."
"Invasion on: Warships Shell Le Havre as 'Chutists Drop from Sky; Two Cities Bombed," *Tulsa Daily World* (June 6, 1944).
"Invasion Thrust Only—Prelude to Coming Battle for Europe"; "Hitler's Errors and Grim Allied Effort Formed Background for Invasion Surge," *Newsweek*, Supplement to issue of June 12, 1944, Vol. XXIII, No. 24, 7; pictures and maps, *ibid.*
Lindley, Ernest K. "High Strategy and the Timing of the Second Front," *ibid.*; "The General on D-Day," *ibid.*, 7–8; "The Men on D-Day," *ibid.*
"Officials Clear Up Details for Feb. 14–16 Draft," *Tulsa Daily World* (January 28, 1942).

"U. S. Declares War . . . ," *Tulsa Tribune*, Vol. XXXVIII, No. 62 (December 8, 1941).

White, James D. "Log-Cabin Diplomacy Deals Axis Another Jolt." Map titled: "Invasion of Africa Puts New Area in War Spotlight." *Tulsa Daily World* (Sunday, November 22, 1942), Sec. 5, p. 8.

"World at War: Every Continent—Every Ocean Is Involved in This Battle to Keep Freedom for All," *Tulsa World* (Sunday, December 28, 1941), Sec. 5.

Index

The paper on which this book is printed bears the watermark of the University of Oklahoma Press and has an effective life of at least three hundred years.